THE

INVISIBLE

SKY

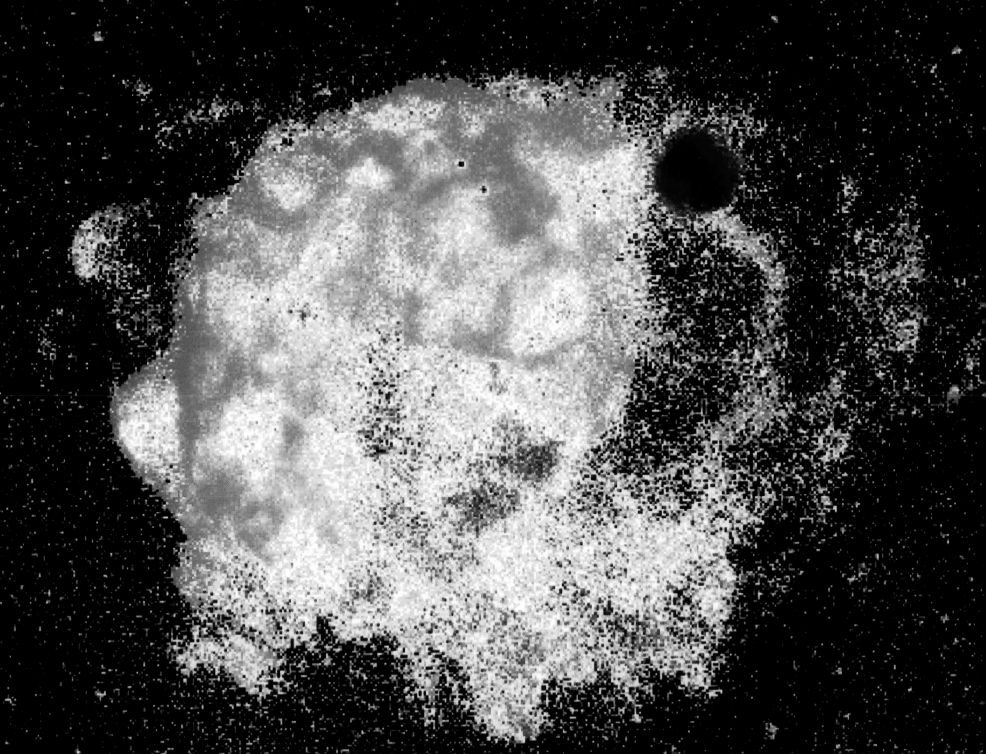

THE INVISIBLE SKY

ROSAT AND THE AGE OF X-RAY ASTRONOMY

BERND ASCHENBACH • HERMANN-MICHAEL HAHN • JOACHIM TRÜMPER
TRANSLATED BY HELMUT JENKNER

COPERNICUS
AN IMPRINT OF SPRINGER-VERLAG

Published in the United States by Copernicus, an imprint of Springer-Verlag New York, Inc.

Copernicus
Springer-Verlag New York, Inc.
175 Fifth Avenue
New York, NY 10010

Library of Congress Cataloging-in-Publication Data

Aschenbach, B.
    The invisible sky : ROSAT and the age of x-ray astronomy / Bernd Aschenbach, Hermann-Michael Hahn, Joachim Trümper ; [translated by Helmut Jenkner].
        p.   cm.
    ISBN 0-387-94928-3 (alk. paper)
    1. X-ray astronomy.   2. ROSAT (Artificial satellite)   I. Hahn, Hermann-Michael.   II. Trümper, J. (Joachim), 1933–   .   III. Title.
    QB472.A83   1993
    522′.6883—dc21   97-34139
                                                                    CIP

Manufactured in Italy.

Printed on acid-free paper.

9 8 7 6 5 4 3 2 1

ISBN 0-387-94928-3   SPIN 10557928

Table of Contents

X-Ray Astronomy Outside Our Galaxy

Preface

Astronomy is not confined to the exploration of the visible sky: Since the fifties, scientists have opened more and more new windows to the universe, making it possible to study numerous new aspects of cosmic events.

The German science satellite ROSAT, circling Earth since June 1, 1990, is an important milestone on this road. Its data have provided us not only with a complete survey of the x-ray sky in several colors, but also with important insights into normal and exotic cosmic objects. It can be said without exaggeration that the entirety of these data has changed our view of the world in which we live.

Soon after launch, we wanted to make the results of our astronomy with ROSAT available to the public. But requests from the publisher had to be put off for quite some time. For scientists at the Max Planck Institute, research and the publication of the results in scientific journals must take precedence. That the thoughts and ideas finally resulted in an actual book is due to the efforts of the science journalist among us, who sifted through the rich harvest of ROSAT data for several months and prepared this book in continuous dialogue with the rest of us. Of course, we had to develop some "courage of omission" and concentrate on those areas that can be conveyed without substantial prerequisites; but we have tried to take into account all crucial aspects and have striven for factual correctness. We also used this courage of omission when personal contributions of the investigators working with ROSAT were involved. While we mention scientists by name in the historical chapters, we preferred to attribute all ROSAT results to the satellite itself, as it were – otherwise, the list of names would have grown too long, and the danger of forgetting one or the other would have been too high.

We have tried to find words to describe the fantastic world of ROSAT, and to join them with the most intriguing images of the invisible sky, for those who make science projects like the ROSAT x-ray satellite possible with their tax contributions. Our thanks go to them as well as to the individuals providing help and review during the development of this book.

Bernd Aschenbach
Hermann-Michael Hahn
Joachim Trümper
July 1996
Garching, Germany

Introduction

The Invisible Sky

For thousands of years our visual perception has ruled our notions about the world. What we see with our own eyes leaves much deeper impressions in our experience than any other kind of information–perhaps with the exception of pain, but this is mainly a protective function. "A picture is worth a thousand words" is the core principle of modern information and advertising strategies.

But what we see is only a minute portion of reality. Not because our sense of vision is insufficient to probe extremely distant or exceedingly small objects: We have learned to overcome such shortcomings. We can explore distant galaxies with powerful telescopes or send space probes to the planets of our solar system, providing us with data and images of our neighboring worlds, or we can peer into the microcosm using microscopes.

But our eyes can show us only a very limited slice of the world, because visible light carries only a small part of the information about physical reality. Other facets remain hidden.

The German-born astronomer William Frederick Herschel showed this fundamental limitation of our vision for the first time in 1800. He detected invisible forms of radiation in a simple but convincing experiment. Using a glass prism, he scattered the light of the sun into the colors of the rainbow and placed thermometers at the various colors in order to measure the intensity of the corresponding radiation. A thermometer placed outside the band of colors, beyond the red end, also showed an increase in temperature. Herschel had detected infrared radiation. Heinrich Hertz

achieved the second "border crossing" in Bonn, Germany, in 1888 when he generated and detected radio waves. The Scottish physicist James Clerk Maxwell had predicted their existence as part of his theory of electromagnetic radiation more than a quarter century earlier. Finally, in 1895, Wilhelm Konrad Röntgen detected a strange new kind of radiation that could penetrate matter. These mysterious rays, which he termed "x"-rays could even disclose details, especially bones, inside the human body.

Soon it became obvious that Röntgen's x-rays–unlike Herschel's infrared and Hertz's radio waves–were located far beyond the violet end of the spectrum. Visible light, as another consequence of Maxwell's theory, was only a small part of the electromagnetic spectrum, comparable to just a single octave of a much wider keyboard.

These discoveries had considerably broadened the horizon for physicists within barely a century, but for some time they had little direct influence on astronomy. Earth's atmosphere does not permit radiation other than visible light and radio waves to reach the ground without being fully or partly absorbed–fortunately for all living beings (including astronomers), since the unweakened ultraviolet, which immediately borders the violet end of the visible spectrum, poses a serious threat to life on Earth.

Although stargazers had always used the "optical window," it took astronomers until the middle of the twentieth century to recognize and exploit the much wider "radio window" in the atmosphere. Although the American astronomer Karl Guthe Jansky discovered the "static" originating from the Milky Way in the early nineteen thirties, detector

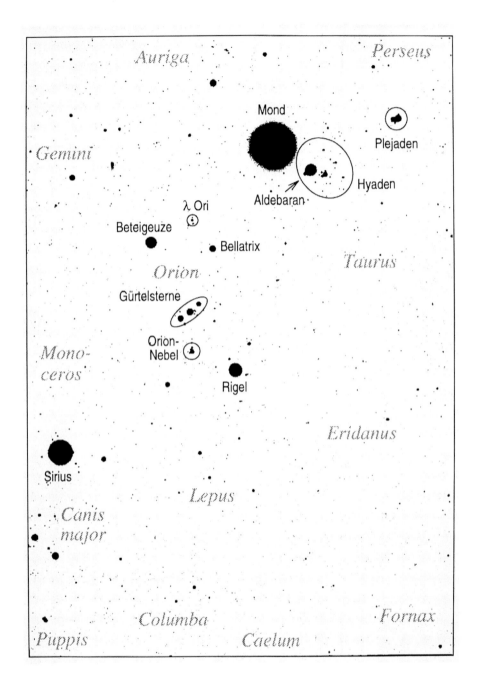

A view of the visible winter sky. The size of the symbols is a measure of the brightness of the objects in visible light; therefore, the moon as the brightest object appears very large (see also the picture on the opposite page).

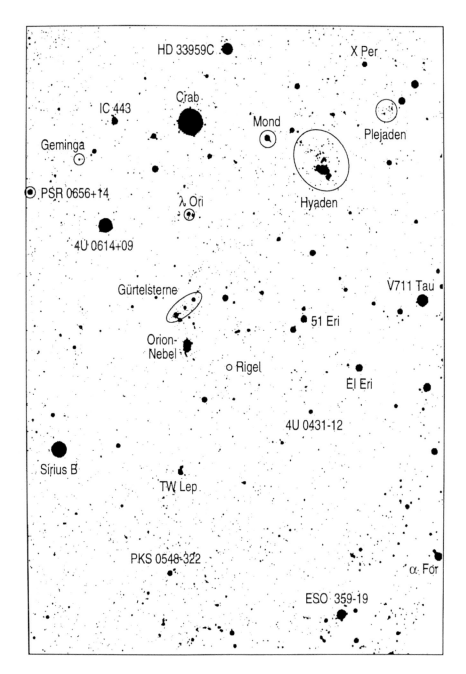

A view of the invisible winter sky. The size of the symbols is a measure of the x-ray brightness of the objects; the most prominent sources are the Crab Nebula, Sirius B, and several x-ray binaries (see also the picture on the opposite page).

and amplifier technology could not then support detailed scientific exploration of the weak cosmic signals. But the development of radar technology during World War II enabled radio astronomy to contribute significantly to the exploration and understanding of the universe.

Other spectral ranges–with the exception of a few very narrow regions of the infrared–can be received only by leaving the absorbing atmosphere behind, by placing telescopes and detectors on high-altitude balloons, sounding rockets, or artificial satellites. Military goals furthered developments in this area as well.

The step into space opened up a new view of the sky, invisible to our eyes and significantly different from the ordinary starry night. While our grandparents experienced the world of stars as infinite eternity, the x-ray sky presents itself as the stage for perpetual fireworks in ever-changing patterns. The differences could not be more striking. Extreme environmental conditions are required for the formation of x-rays: Since x-rays have much shorter wavelengths–and are therefore more energetic–than visible light, considerably more energy is required to generate them.

An example may illustrate the magnitude of this difference. Our sun is the brighest object in the sky. It emits not only visible light but also (with less intensity) x-rays, as discovered by a sounding rocket in 1949. The visible light from our sun originates from a layer only a few hundred kilometers thick, called the photosphere. This layer represents the surface of the sun–we cannot look any deeper, as the gases become hotter and therefore opaque. Because the sun's surface has a temperature of about 5800 K, it cannot produce x-rays, which require several million to a billion degrees. Such conditions can be found far above the surface, where the gases are much more rarefied. This region, the corona, has a temperature of several hundred thousand to a few million degrees–enough for the gases to emit x-rays.

In addition to these so-called "thermal" x-rays, which are created solely by the high temperature of matter, other energy sources can also create x-rays. Electrically charged particles, for instance, have to radiate energy as they change direction on their course through the universe. The more energetic the particles and the sharper the change in direction–caused by cosmic magnetic fields–the larger their loss of energy. Therefore, this "synchrotron radiation" always indicates either extremely energetic particles or very strong magnetic fields, or both.

Outside the range of visible light–and therefore mostly outside the atmosphere–a stormy universe presents itself to the observer. From x-ray astronomy we have learned of many cosmic objects dominated by very high temperatures, extremely energetic particles, or super-strong magnetic fields. Exploring these phenomena requires the appropriate detectors.

The restriction of our natural vision to visible light is a result of the economical evolution principle: In no other spectral range could we perceive our surroundings as brightly. Our atmosphere absorbs other forms of electromagnetic radiation, and the sun emits the largest part of its energy in the visible range. In a sense, we have to count ourselves lucky. It is difficult to imagine a scenario where the sun would emit radio waves, for which the atmosphere is also transparent, so that instead of two eyes we would have to deal with two un-wieldy antennae – not to mention that evolution may have been overtaxed with the development of a radio receiver as a "visual sense."

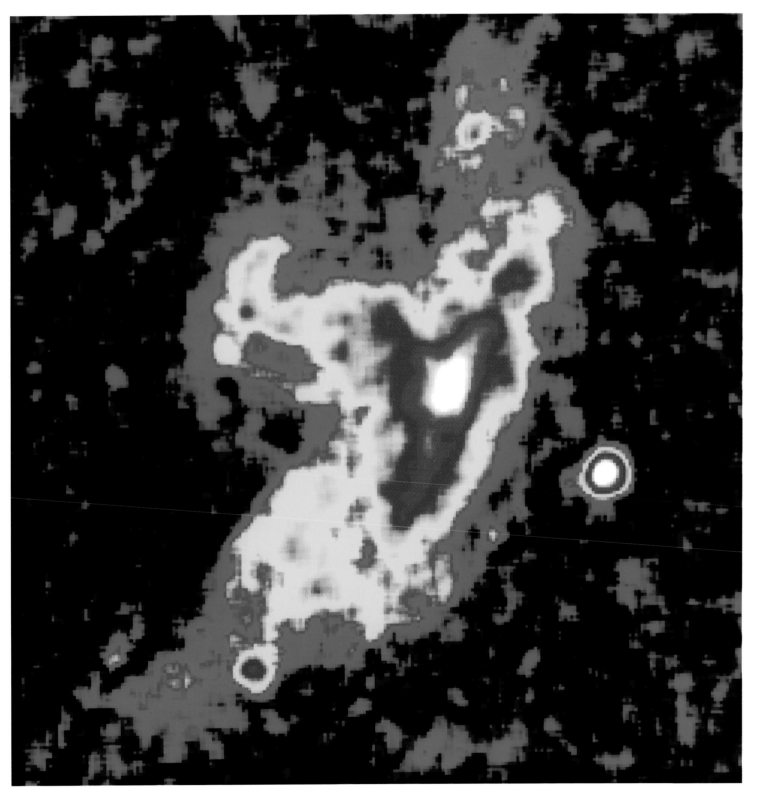

The supernova responsible for this remnant (cataloged as G166.0+4.3) exploded in an environment with relatively high density of matter. Toward the west (right in the picture) the density drops off rapidly, and therefore the shock front of the explosion exhibits a much larger radius, manifesting itself in the form of wing-like extensions toward north and southeast.

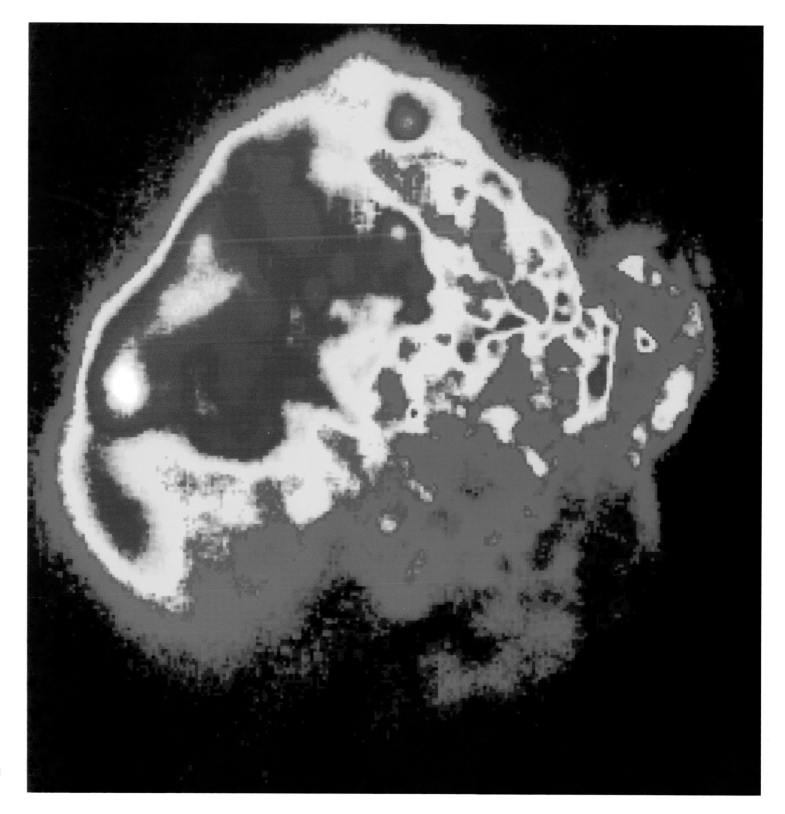

Puppis A is a 3000- to 4000-year-old supernova remnant. In its center ROSAT detected a neutron star with a temperature of several million degrees (central red dot).

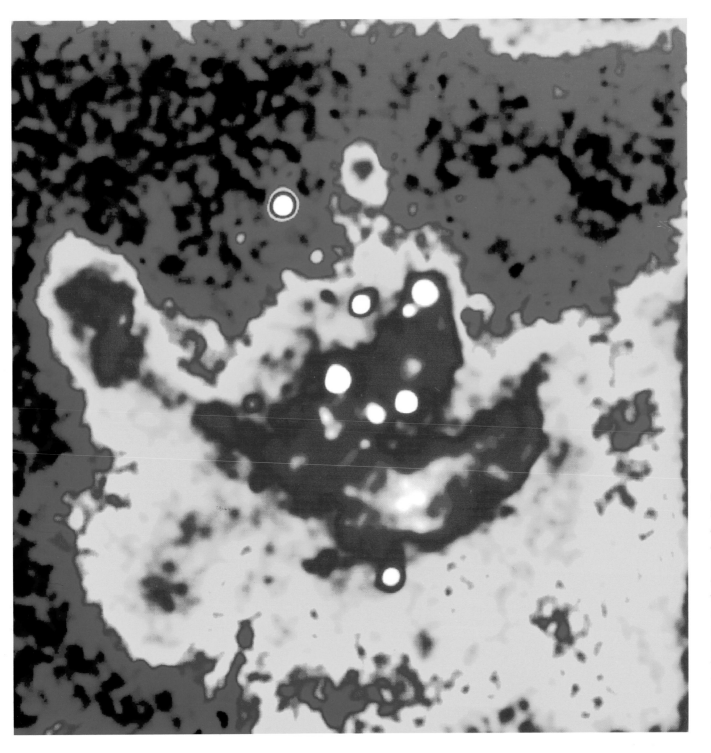

ROSAT image of the Carina Nebula in the Southern sky, containing one of the most massive associations of hot O-, B-, and Wolf-Rayet-stars (white dots) in our galaxy. The gas of several million degrees temperature (yellow and red) fills a volume of almost 300 light years diameter, and originated either in fast, dense stellar winds or from supernova explosions. The variable star Eta Carinae with its surrounding Homunculus Nebula appears as a white area below the center of the image.

The History of X-Ray Astronomy

In the 1870s physics seemed on the verge of being able to describe the world in its entirety. Mechanics allowed the calculation of the behavior of solid bodies; thermodynamics showed the transformation and interaction of different aggregate states and made it possible to determine energy balances. Finally, scientists could explain the electric and magnetic forces and their consequences on matter using electrodynamics.

In 1874, a young high school graduate asked the Munich physics professor Philipp von Jolly whether his A in mathematics would qualify him to study physics. The response: "There may well be the need to explore and classify a little speck here or a bubble there, tucked away in a dark corner, but the system as a whole is well established, and theoretical physics approaches the degree of completeness that geometry has possessed for centuries." Luckily, the young student was not dissuaded–he was Max Planck, who would later receive the Nobel Prize in physics.

This complacency may have been common among physicists of that time, bit it was not long before a completely new, invisible world shattered their expectations. The discovery of electromagnetic waves by Heinrich Hertz in Bonn in 1888 could be viewed as experimental confirmation of electrodynamics, and therefore not necessarily as a novelty. But the reports coming from Würzburg early in 1896 caused quite a stir among scientists and the general public, and led to a period of rethinking and relearning. Röntgen reported the discovery of mysterious rays that penetrated even thin metal plates and could show the position and size of bones in the human body. As Max von Laue showed about fifteen years later with his

Wilhelm Konrad Röntgen (1845–1923) discovered the mysterious x-rays in 1895. (Source: German Museum, Munich.)

diffraction experiments, Röntgen had discovered a new variant of electromagnetic radiation of much shorter wavelength – and therefore of higher energy – than visible light, so that these rays were able to "squeeze" even through thin metal plates.

In the same year of 1896, in Paris, Antoine Henri Becquerel detected yet another enigmatic, even more energetic radiation, to which even bones posed no serious obstacle: it could be absorbed only by thick lead plates. Soon it was shown that this radiation consisted of tiny bullets, small particles ejected from the interiors of atoms (atomic nuclei, about which very little was known at that time). Becquerel had discovered the radioactive decay of uranium and the resulting gamma radiation.

To gain quantitative evidence of this radiation, Hans Geiger developed a device in 1913 to measure radioactivity. Essentially the same device is used to detect cosmic x-rays today. The original detector consisted of an air-tight, gas-filled metal cylinder with a thin wire tip protruding into it. A sufficiently high voltage (several hundred to a thousand volts) is established between the tip and cylinder, creating an electric field inside the cylinder. As an energetic particle of radioactive radiation enters the measuring device, it collides there with one of more atoms. As this happens, one or two electrons may be knocked from the outer shells of the atom. The electric field then functions as a sorting device. The negatively charged electrons are attracted by the positively charged metal tip, and they ionize more atoms on their path towards it; the process repeats itself, so that in the end, an avalanche of electrons migrates towards the metal tip. Meanwhile, the remaining positively charged atoms, the ions, move toward the negatively charged walls of the cylinder. In combination, a measurable current is produced. As long as the applied voltage is not too high, such a device not only registers the presence of an energetic particle; it can also measure its energy to some degree. The current is directly proportional to the absorbed energy, hence the name "proportional counter." Above a certain voltage, however, this additional information is lost, as each incoming particle gives rise to the same current.

It was soon shown that such counters could also be used to detect x-rays. Unlike x-ray emulsions, they did not provide any images of shadowing structures, but in some applications there was nothing to be screened, and only the proof of existence was of interest.

An experiment at the Eiffel tower in Paris in 1910 showed the existence of an ionizing radiation of unknown origin at high altitudes. A strong source of radioactive radiation was placed at the base of the 1000-foot-high structure, and its ionizing effect was measured at the top of the tower and at ground level at a distance of 1000 feet. The measuring devices indicated significantly lower signals on the ground compared to those 1000 feet up. The Austrian Victor Franz Hess carried out even more impressive measurements during balloon flights at altitudes up to 15,000 feet. In 1912, he declared that radiation "of very high penetrating power, entering the atmosphere from above" was responsible for the ionization of the atmosphere.

This radiation proved to be much more penetrating than x-ray or gamma radiation, and during the following years it became clear that it was an extremely energetic form of particle radiation, the so-called "cosmic radiation," consisting mainly of protons but also of heavy atomic nuclei and electrons.

During the 1930s, the French scientist Bernard Lyot collected, unconsciously at first, evidence for the existence of "natural x-ray sources" in the universe. He had developed a method to investigate the spectrum of the solar corona without time pressure. Previously, astronomers had had to rely on the few hectic minutes during a total solar eclipse for their studies of the outer layers of the sun. Using a special filter he developed, Lyot was able to reduce the total flood of light from the sun to a manageable quantity within a minute spectral range, making it possible to target specific spectral lines. Each spectral line is related to a different chemical element, and studying these lines allows a degree of remote chemical analysis. Since the appearance of the spectral lines also depends on physical parameters like density and temperature of the gas and the speed of the source relative to the observer, they also provide additional information about the state of matter.

Evidence of previously unknown elements had been discovered in the solar spectrum during the second half of the nineteenth century, following the development of spectral analysis by Gustav Robert Kirchhoff and Robert Wilhelm Bunsen. One of these new elements, also detected on Earth in 1896, was called helium (after Helios, the Greek god of the sun). Another "solar element," however, evaded further earthly identification, and was known to experts in the field only as "coronium."

Upon measuring the profiles of coronal lines, Bernard Lyot was surprised to find that they were extremely broad, much broader than comparable lines in the chromosphere, the sun's inner atmosphere. This could only be explained by high gas pressure or high temperature in the corona. Both seemed physically unlikely at the time. How could the pressure and temperature of an atmosphere increase toward the outside when the heat source lies within the sun and the outer corona borders cold space?

Even before Lyot's mysterious observations, astronomers had puzzled about the sheer size of the solar corona. It could be traced over several solar radii, or several million miles, during a total eclipse. From the intensity distribution within the corona they determined the decrease of density as a function of distance from the sun's surface and found that the density reaches one percent of the value at the surface only after 400,000 miles; in the photosphere further in, density decreases about 400 times faster.

Three parameters govern the structure and layering of an atmosphere: the gravity of the central body, the mass of the atmosphere's atoms and molecules, and the temperature. The stronger the gravitational pull from the central body, the tighter the layers of the atmosphere will be packed, and heavier elements will always tend to be lower than lighter ones. Independently, a hot atmosphere will be thinner than a cold one – this is also the principle exploited by the sport of hot air ballooning, as the hot air inside the balloon is thinner (and therefore lighter) than the surrounding air, providing the necessary lift.

While the gravitational pull of the sun decreases to about one-fourth of its surface value at 400,000 miles from the surface, this alone cannot explain the observed decline of the corona's density. There is still a discrepancy of about hundred times, which can only be explained by a corresponding increase in temperature.

Does this indicate that the solar corona indeed has a temperature of several hundred thousand to a few million degrees? In 1942, the Swedish astronomer Bengt Edlén provided another clue. He showed that the mysterious "coronium" lines, as well as several other lines in the solar spectrum that had not been understood, did not come from a new species of atoms but from atoms that had lost the majority of their electrons due to the extremely high ambient temperature and that therefore are strongly electrically charged, or ionized. However, these electrons cannot move around freely. They are attracted by the strongly ionized atoms, and even captured at times; during this process, energy can be emitted as radiation.

X-Rays from the Sun?

It is not too surprising that during the mid-forties, as research returned to normal after the end of World War II, scientists became extremely interested in determining the temperature of the solar corona by other means. A key experiment would be the detection of x-rays from the corona, since a temperature of the order of a million degrees should produce soft, or relatively long-wavelength, x-rays.

American scientists began to exploit the V-2 rockets, which had been seized in Germany shortly before the end of the war; as experience with this technology was scarce, a number of German rocket scientists and engineers had been brought back to the United States as well.

The first attempt to detect solar x-rays occurred in 1948. A V-2 rocket, converted for scientific use, carried a number of photographic plates; thin beryllium foils prevented the plates from exposure to visible light, but they did not block the more energetic ultraviolet and x-ray radiation. The rocket reached an altitude of about 100 miles, returning to the ground with the help of a parachute. When developed, the photographic plates showed clear signs of exposure. However, the source could not be identified, as the radiation could have arrived at the plate from any direction.

A year later, Herbert Friedman, of the Naval Research Laboratory in Washington, DC, used a number of Geiger counters as payload on another V-2 rocket. In addition to simply detecting the radiation, this experiment also attempted to provide some directional information by using photoelectric cells sensitive to the sun's visible light in addition to the x-ray detectors, and by taking advantage of the rocket's rotation around its axis, required to achieve stability during flight. As the x-rays from a given source appear strongest as they pass through the field of view of the measuring aperture, their intensity would vary with the rotational period of the rocket. A simple comparison of the x-ray variations with the signal from the photo-cells then proved that the x-rays indeed originated in the direction of the sun.

This was a gratifying result for solar physicists, since they now had reliable proof of temperatures of at least a million degrees in the solar corona. It also confirmed that the radiation from the

Riccardo Giacconi, one of the pioneers of x-ray astronomy. He discovered the first cosmic x-ray source outside the solar system.

corona causes the existence of Earth's ionosphere, particularly the so-called E-layer, which is of great importance in global radio communication.

However, the prospects for detecting x-rays from other sources appeared very dim indeed: The closest star is 300,000 times more distant than the sun, and so its radiation would appear 90 billion times fainter. Such measurements, given the x-ray detectors at that time, were unthinkable. X-rays from the center of the Milky Way or from other galaxies seemed forever inaccessible.

A Fortuitous Discovery

During the nineteen fifties, the exploration of cosmic x-ray radiation was only a marginal sideshow to the exciting discoveries at the other end of the electromagnetic spectrum, the radio waves. But the success of the "competition" also stimulated a few scientists who tried to achieve the seemingly impossible. One of them was Riccardo Giacconi, who had studied physics in Milan, Italy, gone to the United States to pursue particle physics at Indiana University and Princeton, and then explored the collisions of cosmic ray particles with the atoms of Earth's atmosphere. In 1959 he left the university and joined a group led by Bruno Rossi, who had advanced research in the area of cosmic radiation during the forties, taught at the Massachusetts Institute of Technology (MIT), and now worked as an advisor for the American Science and Engineering Company (AS&E). Rossi understood the possibilities offered by the young field of space flight in the quest to explore the universe. He encouraged Giacconi to join in the search for x-rays from space.

Getting funds for such a hopeless undertaking was no easy task. Finally, a secret Air Force lab showed interest in a new attempt to confirm the existence of cosmic x-rays. However, their interest was concentrated on potential x-rays from the moon. Of course, the Air Force knew that the moon is far too cold to radiate x-rays. But the moon is constantly exposed to the solar wind, an endless flood of electrically charged particles evaporating from the sun. As these energetic particles strike the lunar surface, x-rays may be created much as they were in the experiment that led to their discovery by Wilhelm Konrad Röntgen at the end of the nineteenth century. This does not happen on Earth, where the magnetic field shields the surface from the solar wind. The Air Force thought that x-ray radiation from the moon might provide clues to the energy of the solar wind particles.

A second kind of x-ray radiation from the moon was conceivable as well. The moon is constantly exposed to x-rays from the sun, and the released energy might be sufficient to generate new x-rays. As an x-ray photon interacts with an atom's electron, it may temporarily dislodge the latter from its orbit. But as long as it stays close enough to the atom's nucleus, the electron will be recaptured into its original orbit and give off the previously acquired energy. This x-ray fluorescence radiation can even provide information about the chemical composition of the lunar surface. This was also of interest to the Air Force scientists, since President John F. Kennedy had just announced the start of the Apollo program. They wanted to see if the chemistry of the moon could be explored from Earth.

Unfortunately, the first launch attempt, in October 1961, failed. While the rocket flew into the night sky above the test range of White Sands, New Mexico, according to plan, the entrance window in front of the x-ray detectors did not open. Giacconi's group attempted a second launch on June 18, 1962. Again the full moon stood over White Sands, this time in the constellation Sagittarius, not far from the center of the Milky Way. This time, everything went according to plan, and after a few minutes, two of the three Geiger counters reported stong x-ray radiation.

However, analysis of the data showed that the radiation could not have come from either the sun or the moon: The strongest signal was measured by the detectors when they were pointed toward the constellation Scorpius, about forty degrees from the position of the moon; a second, significantly weaker source appeared in the direction Cygnus.

Filling up the Sky

As Giacconi and his collaborators began to search optical sky photographs for indications of peculiar objects, their early euphoria of discovery turned into disappointment. In the area of the sky in question they could find nothing that could be easily identified as an x-ray source. So they restricted themselves to a simple nomenclature containing the name of the constellation and a number: they named the first x-ray source Scorpius X-1. The reaction of astronomers to Giacconi's discovery was muted.

In the following years, several more rocket flights were carried out. On April 29, 1963, a group at the Naval Research Laboratory under Stuart

Bowyer found a source in the constellation Taurus (Taurus X-1). This time the source was not hard to identify. In this area of the sky, optical astronomers could provide a special object: the Crab Nebula, remnant of the famous supernova of 1054.

The assumed connection between an x-ray source and the remnant of a stellar explosion provided astronomers with a first clue toward the explanation of at least one object. Herbert Friedman, who had already distinguished himself in the exploration of x-rays from the sun, pointed out that in the thirties Fritz Zwicky and Walter Baade had attempted to lay the theoretical foundations for the last stages of evolution of very massive stars. They had concluded that the core of a massive star collapses into an extremely compact object while the outer envelope is ejected in a huge explosion. The two astrophysicists associated this explosion with supernovae, whose gigantic energy bursts were then completely mysterious.

But in such a compact object, matter had to have lost its normal structure. It could not consist of ordinary atoms, since they are essentially "empty" and cannot be packed densely enough: Electrons (carrying negative electrical charge) circle the nucleus, consisting of (positively charged) protons and (electrically neutral) neutrons, at a distance of many thousand times the diameter of the nucleus. The extreme pressure in the center of a supernova would press the electrons into the nucleus, where they would combine with the protons to form neutrons. The core would become a mush of neutrons, according to Baade and Zwicky. Because neutrons do not repel each other electrically, they could be packed extremely densely, up to a limit set by quantum physics. A neutron star could therefore contain, in a diameter of only a few dozen miles, more mass than our sun.

Friedman now pointed out that such a compact neutron star would have to have a surface temperature of several million degrees and could therefore qualify as an x-ray source. The question remained whether this "thermal" x-ray radiation would be strong enough to be registered by the detectors of the 1960s over a distance of several thousand light years.

While it was not possible for the time being to identify the x-ray source in Taurus as a neutron star, the correspondence of supernovae and x-ray sources proposed by Friedman was quickly accepted. The lack of angular resolution of x-ray detectors, however, provided a major impediment to understanding x-ray sources and finding their optical counterparts.

The Moon as Observing Assistant

Nature soon came to the rescue. Taurus X-1 lies close to the ecliptic, the celestial "highway" on which sun, moon, and planets appear to run. On July 7, 1964, the moon moved in front of the Crab Nebula and served as a huge light pointer in the sky: It would suffice to observe the source while it was being occulted by the moon, and to note the time when the x-rays were eclipsed by it – then the source would be located just behind the leading edge of the moon's disk. In this way, not only could the exact correspondence of the Crab Nebula and the x-ray source be confirmed, but its diameter could be measured as well. The x-ray signal from the source did not disappear instantaneously (as would have been expected if it were a neutron

The Crab Nebula is the remnant of a supernova explosion, observed on July 4, 1054, from many places on Earth and documented by Chinese historians.

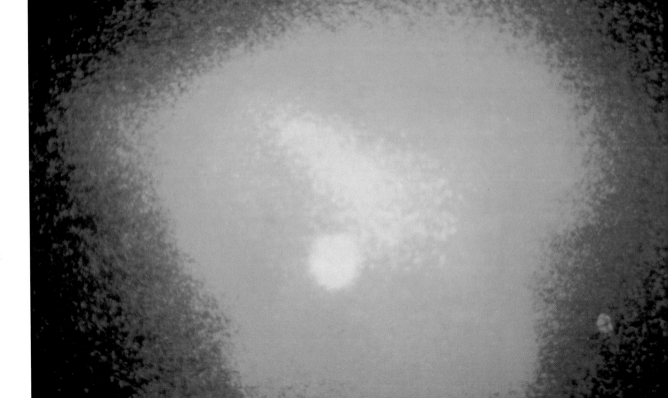

X-ray image of the Crab Nebula, obtained by the American HEAO-2 satellite (Einstein); a bell-shaped area of radiation is visible around the pulsar, appearing particularly bright in a small strip.

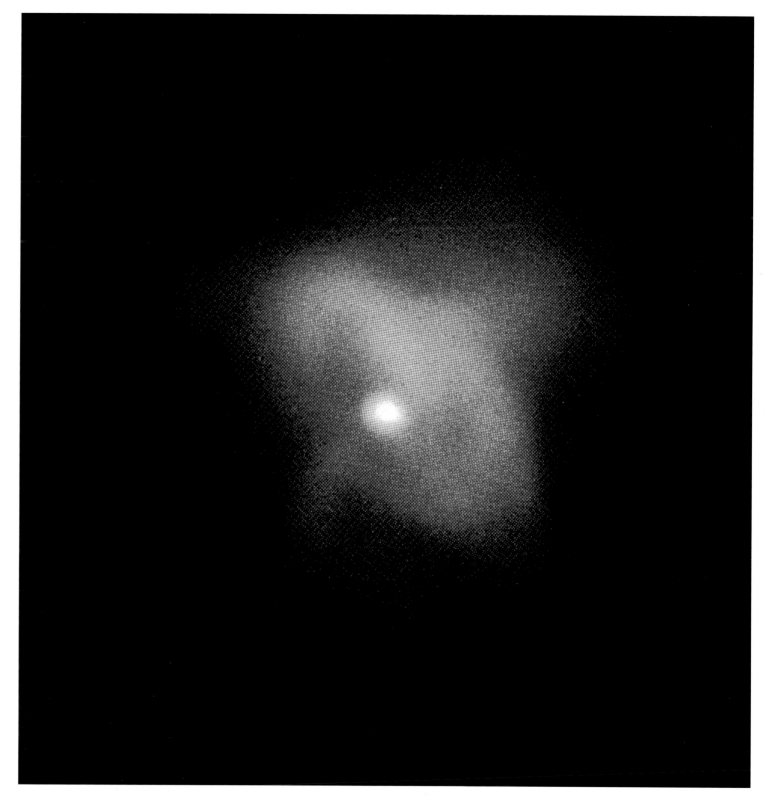

The ROSAT image of the Crab
Nebula is rich in detail and resolves
the bell-shaped area of radiation
into a ring-shaped structure around
the equator of the neutron star and
into a jet along its rotation axis.

star) but faded over a period of about two minutes. Therefore, the x-ray source Taurus X-1 must have an angular diameter of approximately one arc minute (one-sixtieth of a degree, or one-thirtieth the angular diameter of the moon), corresponding to a physical diameter of about two light years, given the assumed distance of about 6500 light years.

Rotating Blinders

For the remainder of the then known x-ray sources, the services of the moon were unavailable, physicists had to find other ways to fix their positions. As a first step, they placed blinders of sorts on their detectors, and they put them into a controlled rotation. As blinders they used collimators which resemble a grate of wide vertical metal slats. Radiation entering through the individual openings could fall on the detectors only if the window was fairly accurately pointed at the source. This technique reduced the area of sky covered by the detector dramatically. In addition, it improved not only the positional resolution of the device but also its sensitivy, as disturbing radiation from neighboring areas of the sky was eliminated.

Controlled rotation improved the resolution even more, since the maximum signal appeared only if the source was located exactly in the center of the restricted field of view. Scans of the same area of the sky in different directions would yield different curves of measurements. The peaks of those curves agreed with each other only if the scans crossed at the position of the source. In addition, the rotating detector system could identify multiple sources within a single field of view.

Such measurements, however, took much longer than the few minutes that were available during the flight of a sounding rocket. This increased the desire among x-ray astronomers for experiments from orbiting satellites, which would allow much longer observations. When Giacconi became a member of a panel on space science of the National Academy of Sciences in 1965, he generated a report to the National Aeronautics and Space Administration (NASA) on the state of x-ray astronomy with the support of Rossi and Friedman. It stated that the exploration of cosmic x-rays would play a significant role in the effort to achieve a better understanding of the universe. A year later, Giacconi became project leader for SAS-A (Small Astronomy Satellite A), which was to be developed and flown as part of the Explorer program.

With Uhuru Toward New Frontiers

Two proportional counters filled with argon gas formed the heart of this first x-ray satellite; their sensitivity was more than thirty times higher than any previously used detector. This sensitivity also made them vulnerable to hits by energetic particles from the Van Allen radiation belt, which encircles Earth and comes fairly close to the surface above the South Atlantic. These belts were detected in 1958 by astronomer James Van Allen during an experiment on the first U.S. satellite, Explorer I. To avoid the Van Allen belt as much as possible, a low Earth orbit of about 350 miles altitude, close to the equator, was selected. From

The American x-ray satellite Uhuru produced the first survey of the x-ray sky during the early seventies.

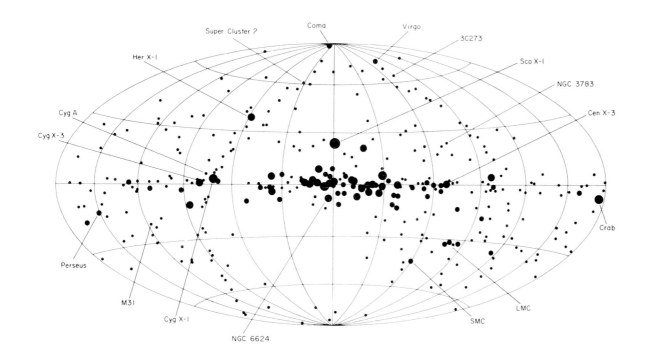

The American x-ray satellite Uhuru found about 400 x-ray sources. The objects along the center line (the plane of the Milky Way) belong to our galaxy, while most of the sources away from the plane are linked to other galaxies.

the U.S., such an orbit could be achieved only with additional course corrections, since the latitude of the launch site automatically determines the initial orbit inclination. The intended launcher, a Scout rocket, could not perform this maneuver.

To overcome this obstacle, NASA entered into an agreement with the space flight center of the University of Rome to launch the satellite from the Italian launch site of San Marco, off the coast of Kenya, only three degrees south of the equator. Not only could the launch exploit the full momentum of Earth's rotation, but no additional course correction would be necessary. Launch took place on December 12, 1970, the seventh anniversary of the independence of Kenya. To honor this date, SAS-A received a new name: Uhuru, the Swahili word for freedom.

Collimators restricted the two measuring windows of the proportional counters to five by five degrees and to five by one-half degrees respectively. The satellite rotated with a period of twelve minutes, scanning a strip of sky at every rotation; the direction of the satellite axis (and therefore the position of the measured strip of sky) was kept constant for 24 hours. Then, an electromagnetic torquer aboard the satellite was switched on, serving as a kind of "hook" for Earth's magnetic field, so that the direction of the rotational axis (and therefore the position of the observed sky strip) could be changed by a small amount.

In this way, almost the entire sky was surveyed for new x-ray sources during the first few months of the mission. One of the biggest surprises was the detection of extremely rapid variations in the x-ray source Cygnus X-1.

Cygnus Cycles

The source in the constellation Cygnus had been detected during Giacconi's first x-ray measurements in June 1962 and had attracted attention during later sporadic observations by its changing intensity. But now the Uhuru results showed that the x-ray intensity of Cygnus X-1 changed significantly over a few seconds.

Since the satellite rotated slowly, each source remained in the field of view of the detectors for about twenty seconds, and with individual "exposure times" of 0.4 seconds, corresponding short-term variations could be detected. The signal from Cygnus X-1 showed jumps of fifty percent or more within fractions of seconds.

Interestingly enough, the individual intensity peaks did not follow a fixed periodic pattern, and so the scientists reduced the detector "exposure times" to search for even shorter variations. Even at 0.1 seconds, no regular pattern of intensity variations appeared. At first, a period of 73 milliseconds seemed to be indicated, but additional measurements did not confirm these findings. What could be seen, though, were strong intensity variations of the source with times scales from milliseconds to years. From this it followed that the x-ray source had to be an extremely compact object.

The key to understanding the nature of this object was handed to the astrophysicists with the optical identification of a blue giant star that formed a double star system with the compact x-ray source. From the orbital period of 5.6 days, from optically derived orbital parameters, and from the mass of the giant star, the mass of the x-ray star could be derived using Kepler's third law. The resulting value was six solar masses – too much to be a neutron star. Therefore, Cygnus X-1 became the first black hole candidate.

During its aproximately two-year mission, Uhuru identified about 400 x-ray sources, a better than tenfold increase in the number of known sources. Most of the bright sources were concentrated along the plane of the Milky Way, or galactic disk; they could be considered comparative neighbors. A smaller fraction of mostly faint sources was distributed equally over the entire sky. These were classified as extragalactic sources and could be partly identified with clusters of galaxies and quasars.

The positional accuracy of Uhuru's detectors, one to two arc minutes (comparable with the resolution of the human eye), made the identification of most x-ray sources with optical counterparts easier. Identification was particularly simple and rapid with a class of objects where the x-ray signal discontinued for brief periods with constant regularity; this effect could be seen early on in about a dozen of the variable Uhuru sources.

Optical astronomers were familiar with similar behavior in a certain class of variable stars. Not all stars in the sky emit light with constant intensity – even with the naked eye, medium- to long-term variations can be detected. The best example is the star Algol in the constellation Perseus, visible in the northern fall and winter skies: Its brightness drops by more than a magnitude precisely every 2.867 days. The English astronomer John Goodricke was able to provide an explanation for this effect as early as 1782. Algol consists of two close stars of different brightness, circling each

other in such a way that the fainter one crosses in front of the brighter star every 2.867 days and so eclipses it. During this stellar eclipse we see only the light from the fainter star, as opposed to the combined light from both stars we see at all other times. Astronomers call these systems eclipsing binary stars.

A similar explanation offers itself for the transient and regular cessation of the x-ray signal: If one of the stars in such a binary system is an x-ray source, its radiation will be blocked by the other star at regular intervals. If the other star also shows variation with the identical period in visible light, it is easy to identify them as counterparts.

In this way the newly discovered x-ray source Hercules X-1 could be identified with a known variable star – HZ Herculis, a star that drew the attention of the German variable star observer Cuno Hoffmeister in 1936: Its brightness varied apparently irregularly by about two magnitudes. At first, Uhuru's detectors registered strong x-ray bursts from Hercules X-1 every 1.2378206 seconds, indicating a correspondingly rapid rotation of the x-ray source. Radio astronomers had previously discovered the neutron stars predicted by Zwicky and Baade in the form of pulsars (sources of radiation emitting pulses in regular intervals), leading to a simple explanation for the rapid x-ray variation. The x-ray source could only be a rapidly rotating neutron star, whose x-ray beam swept over Earth at every revolution like the beam of a cosmic lighthouse.

At the same time, Uhuru determined that the x-ray source disappeared every 1.7 days for a few hours, and that the 1.24-second rotational period was slightly variable, by 0.7 milliseconds, over the same time interval. This showed convincingly that the rotating neutron star revolved around a companion. Hoffmeister's and other astronomers' observations could now be unified into a single consistent explanation: Plotted against the x-ray period of 1.7 days, the optical data showed a convincing fit, reaching a maximum at the point when the x-ray source was located in front of HZ Herculis–at that point when we look at the hemisphere of the star that is heated by the x-ray source, and therefore brighter.

Bright and strongly variable x-ray sources turned out to be bizarre binary systems where one component had already reached the final stages of stellar evolution, circling its partner as a compact object.

Magnetic Remote Sensing

In the mid-seventies the model for the explanation of variable x-ray sources was impressively confirmed by a rather sensational observation. Scientists from the Max Planck Institute for Extraterrestrial Physics (MPE) in Garching, Germany, together with colleagues from the University of Tübingen, repeatedly sent x-ray detectors into the stratosphere using high-flying balloons at an altitude of 25 miles in order to measure high-energy x-radiation (about 20 kiloelectronvolt), which can penetrate even into these denser layers of Earth's atmosphere. The electronvolt is the unit of energy in high-energy physics; one electronvolt (eV) corresponds to a wavelength of 12,400 angstroms, or 1240 nanometers (billionths of a meter). Since energy and wavelength are inversely proportional (that is, the shorter the wavelength of a given

radiation, the higher its energy), an energy a thousand times higher, or 1 kiloelectronvolt (keV), describes a wavelength a thousand times shorter: 1.24 nanometers. Within high-energy radiation, one distinguishes between "hard" and "soft." The higher the energy, or the shorter the wavelength, the harder the radiation.

The main goal of this long-term observing campaign was the determination of the expected slow decrease of x-ray intensity of the Crab Nebula and its pulsar. At the time, the understanding of pulsars, first detected in 1967, had progressed to the point that the source of the radiation energy was believed to be the rapid rotation of the neutron star: Because of the continuous release of radiative energy, the rotational velocity had to decrease slowly; in other words, the period of the pulses had to increase, an effect radio astronomers had already confirmed. Theoretical investigations had also shown that the x-ray intensity had to decrease with the age of the object, since the decreasing rotational velocity also "paralyzes" the acceleration forces required for the generation of x-rays. This was expected to be fairly easily proved with observation of a comparatively young neutron star like the Crab pulsar.

During an observational flight lasting several hours, other sources could be observed as well, and so the attention of the German x-ray astronomers was also aimed at several of the x-ray binaries detected by Uhuru. Their balloon program, HEXE (high-energy x-ray experiment), did not contain the conventional proportional counter as a central detector, since this device could not measure high-energy x-rays. The gas in the counter's tube would not be dense enough to "stop" and count sufficient x-ray photons. A more appropriate technique applies transparent crystals of sodium iodide or cesium iodide. As a high-energy x-ray photon enters such a crystal, it is absorbed and its energy is transferred to an electron. This leads to ionization and excitation of other atoms, which emit light ("scintillate"). These flashes of light can be registered with photomultipliers, with the brightness of the flashes being a measure of the energy of the absorbed x-ray photon.

Unfortunately, scintillation flashes are also caused by particles of cosmic radiation. To distinguish these "false" signals, a sandwich of a thin sodium iodide and a thick cesium iodide crystal is used with one photomultiplier. The x-ray photons are absorbed in the sodium iodide crystal and generate their flashes there. Particles of cosmic radiation, however, penetrate both crystals and generate a flash in the cesium iodide crystal as well, but with a different signature. This way, the origin of the flashes can be determined and data from the interfering particles discarded. For the HEXE project, the crystal sandwich and its collimator were enclosed in a pot-shaped enclosure of plastic scintillators. A total of sixteen photomultipliers was used.

On HEXE's first flight, on August 28, 1973, from Palestine, Texas, a special balloon for stratospheric flight with a volume of 14 million cubic feet carried the almost 800 pound gondola to an altitude of 26 miles within a few hours. Within the available observing time of 30 hours, the Crab Nebula and several other sources were examined.

In May 1976, the scientists from Garching and Tübingen carried out perhaps the most important measurements of the HEXE program. Again they had set their sights on Hercules X-1, but this time the scientists were less interested in the timing of the x-ray pulses than in their spectrum. They wanted to find out how the intensity of the x-ray radiation changed as a function of energy, or wavelength. In analyzing their data, they found the expected decrease of x-ray intensity toward shorter wavelengths. However, against this trend a small peak appeared around 53 keV. Such a clearly defined intensity peak pointed to a special radiation mechanism.

In visible light, comparable narrow intensity peaks had been known for a long time as spectral lines. Around 1860, Bunsen and Kirchhoff had based their spectral analysis of chemical elements on the investigation of such "emission lines," which originate from electrons jumping from an outer shell in an atom to a closer orbit and releasing in the process their excess energy as radiation in a characteristic wavelength.

As rotating neutron stars, pulsars were known to have extremely strong magnetic fields. As a massive star shrinks to a diameter of only a little more than ten miles, its magnetic field will be compressed very strongly as well–to values of the order or 100 million tesla, or in older units, a trillion gauss (the magnetic field on Earth's surface is roughly half a gauss). To produce the spectral line in Hercules X-1 requires even more extreme conditions. For one, the surface of the neutron star exhibits extremely hot spots of about a million degrees around the magnetic pole caps. Channeled by the strong magnetic field, matter from the companion star HZ Herculis rains down on one of these spots (we cannot see the other one). The extreme gravitational field of the neutron star leads to an infall velocity of about one-third the speed of light, or 60,000 miles per second. Under these conditions about 100 billion tons of matter crash into the polar caps every second. It is no surprise that extremely hot plasma exists there, cooled by strong x-ray emission. The spectral line discovered by the scientists in Garching and Tübingen originates from electrons forced by the super-strong magnetic field into spiral trajectories in this plasma; as in an atom, they can occupy only "discrete" energy states. As such an electron changes from one energy state to another, it emits quantized cyclotron radiation (named after the type of early particle accelerator, where this kind of radiation was first detected) according to the energy difference of this quantum jump; according to the laws of quantum physics, the magnetic field strength alone determines this energy difference. In the case of Hercules X-1 and a measured energy of about 50 keV, this leads to a magnetic field strength of 500 million tesla, or 5 trillion gauss. This was the first direct, that is, spectroscopic, measurement of the polar magnetic field of a neutron star. By 1996, astrophysicists had measured the magnetic fields of about ten neutron stars in this way.

Truly hellish conditions exist at the polar caps of these mass-accreting neutron stars. Not only are the temperatures extremely high and the magnetic fields extraordinarily strong, but the energy density is enormous. The radiation of a speck of surface area with a diameter of about one-thousandth of an inch would suffice to supply the entire annual energy demand of the United States. Each polar cap has an area of approximately half a square mile. These areas lend themselves to studies of physical conditions that are outside the limits of terrestrial laboratories by several orders of magnitude.

Einstein and EXOSAT

In spite of singular successes of balloon-borne x-ray astronomy, a more complete investigation of the invisible sky remained the domain of satellites. During the second half of the nineteen seventies, the United States launched two high-energy astronomical observatories (HEAO) with detectors to observe cosmic x-rays; again, Herbert Friedman and Riccardo Giacconi were in charge of the projects. HEAO-2 in particular led to numerous discoveries during its two-and-a-half-year mission; it was launched in November 1978 and later received the name "Einstein." It delivered x-ray images for the first time; they were obtained with an imaging x-ray telescope 23 inches in diameter; now, details down to about 10 arc seconds could be seen, and numerous x-ray sources showed discernible structure. (See the next chapter for more detail on the principle of x-ray telescopes.)

Not only resolving power increased with the use of such a telescope, but also light-gathering power and sensitivity. In addition, the detectors had been improved. Therefore, HEAO-2 could observe sources that were a thousand times fainter than those accessible to its predecessor, Uhuru. And since resolving power and positional accuracy had reached the arc-second level for the first time, cross-identification with optical sources became increasingly easier. Before HEAO-

In the late seventies the American HEAO-I satellite found about 840 x-ray sources. AGN = active galactic nuclei, SNR = supernova remnants, BINARY = x-ray binaries, CLUSTER = galaxy clusters.

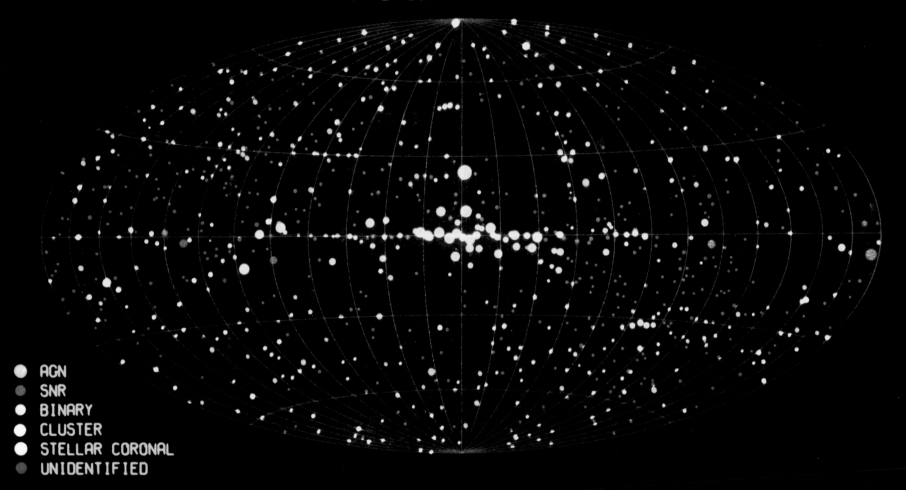

HEAO A-1 ALL-SKY X-RAY CATALOG
NAVAL RESEARCH LABORATORY

AGN
SNR
BINARY
CLUSTER
STELLAR CORONAL
UNIDENTIFIED

2 burned up in the atmosphere in the summer of 1981, scientists had gathered data on more than 5000 cosmic x-ray sources.

Two years later it was the Europeans' turn: On May 26, 1983, the European Space Agency (ESA) launched EXOSAT (for European X-ray Observatory Satellite) into a highly elliptical orbit with an altitude between 220 and 120,000 miles. In this orbit, the satellite took about four days for one revolution of Earth, with only a few hours inside the interfering Van Allen belt; therefore, it could observe x-ray sources with its proportional counters and two 11-inch telescopes without interruption for long periods of time.

Originally, this particular orbit had been selected to make use of the moon as observing assistant: The high inclination of 72 degrees together with the strong eccentricity of the orbit would cause the moon to cover – from the point of view of the satellite – over the course of the mission a fairly wide strip of the sky (from Earth, this strip is only about ten degrees wide), occulting the majority of x-ray sources. On such occasions, the detailed structure and exact position of the sources was supposed to be measured: As an extended source gets occulted by the moon, the x-ray signal would not stop momentarily, but decrease over time, giving information about its structure beyond the resolution of the observing instrument. How-ever, this type of observation requires high time resolution, and the correspondingly short "exposure times" are possible only for very bright x-ray sources.

In the end, this technique was not used; not only because of the above considerations, but also because the engineers were afraid that the sensitive star sensors used to point EXOSAT would be blinded by the moon's bright light.

The group around Joachim Trümper at the Max Planck Institute for Extraterrestrial Physics in Garching was involved in building and calibrating the EXOSAT instruments, and he had proposed to carry out observations of lunar occultations. However, they did not intend to investigate the detailed structure of the sources. Their main goal was the exploration of the faint x-rays that were deflected from their straight paths by interstellar dust clouds. A similar phenomenon is the scattering of visible light from the moon by ice crystals in cirrus clouds, generating a halo.

Unfortunately, these observations could not be carried out. But a large number of astronomers used EXOSAT for many important observations. Its main strength was its four-day orbit, allowing long uninterrupted observations of x-ray sources for the first time. This proved to be particularly fruitful for the studies of x-ray binaries, with their complex intensity variations over time.

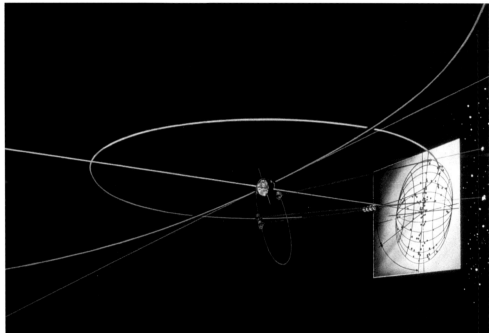

EXOSAT circled Earth every four days in a highly elliptical orbit, allowing continuous long-term observations of variable x-ray sources.

In 1983, Europe launched its first collaborative x-ray satellite, EXOSAT, with two small x-ray telescopes and the largest proportional counter of its time.

ROSAT – Creating a Satellite

X-ray astronomy is a very young branch of celestial research. Few scientists gave it serious thought before the discovery of the first cosmic x-ray sources outside the solar system, in the summer of 1962. No wonder the pioneers in this area – as in radio astronomy a few decades earlier – came mostly from other fields. While radio astronomy had relied on high-frequency engineers and scientists, the pioneers of x-ray astronomy came from nuclear physics or from the study of cosmic rays.

Joachim Trümper, for instance, the father of the ROSAT x-ray satellite, had studied in Halle, Hamburg, and Kiel during the fifties and intended to work as a particle physicist in Hamburg. In search of a subject for his master's thesis in 1957, he came across a group in Hamburg that studied a broad spectrum of topics in nuclear physics under the direction of Erich Bagge. Among these was the study of cosmic rays, discovered by Victor Franz Hess in 1912, which before the development of large particle accelerators served as natural experiments for investigations on elementary particles and atomic nuclei.

Cosmic rays had turned out to consist of extremely energetic particles, bombarding Earth from all directions. The lion's share of them are protons (the atomic nuclei of hydrogen); alpha particles (the nuclei of helium) make up about one-seventh of the total; and nuclei of all heavier elements together comprise about one percent. This distribution corresponds roughly to the average abundance of the individual elements in the universe, making it possible to draw interesting conclusions whenever there is a deviation from these proportions. About three percent of the radiation consists of energetic electrons.

This so-called primary radiation does not reach Earth's surface. Collisions with atomic nuclei in the atmosphere generate numerous "particles," some of which penetrate to the ground: pi mesons, muons (heavy "cousins" of electrons), and positrons (antielectrons). By studying these particles, nuclear physicists hoped to learn about the structure of atomic nuclei.

A New Type of Detector

In Hamburg, Trümper was allowed to help develop the so-called spark chamber, a new type of detector for fast ionizing particles, which would later be used also with large particle accelerators. This new detector, invented by the group in Hamburg, consisted of a number of parallel metal plates inside a gas-filled chamber; these plates are alternately connected to the two poles of a high-voltage power supply. As a fast ionizing particle enters the chamber, it generates electric sparks between adjacent plates along its trajectory. It is sufficient to take pictures of these sparks from two different angles to reconstruct the particle's path. The main advantage of this approach, compared to, say, the bubble chamber, is that the spark chamber can be "switched on" automatically on demand, limiting the measurements to passages of interesting particles.

Trümper and Otto Claus Allkofer used this new detector to make measurements on cosmic rays. On the Zugspitze (Germany's highest mountain) they carried out experiments to determine the energy distribution of muons generated by cosmic radiation.

Next, Trümper and a team of scientists in Kiel developed the "air shower experiment," starting measurements in 1965. This experiment explored particle cascades generated by extremely energetic cosmic rays in Earth's atmosphere: For every single cosmic ray particle hitting the atmosphere, with an energy of one million gigaelectronvolt (GeV, 1 GeV = 1 billion eV), about 100,000 particles arrive at sea level. Their distribution is measured with a large-scale detector array.

The goal of the experiment was to determine the chemical composition of the cosmic radiation at the highest energies. But in spite of a number of interesting results, the physicists of the Kiel group did not achieve that goal. Recently, a new attempt was initiated in the form of the Karlsruhe air shower experiment, or KASCADE (for Karlsruhe Air Shower Core and Array Detector), using significantly more resources and the latest technology.

A Key Experience

Joachim Trümper was interested not only in the chemical composition of the cosmic radiation but in its origin. What kind of object would be capable of accelerating elementary particles or entire atomic nuclei almost to the speed of light, so that they could traverse the entire Galaxy without being slowed down? The magnitude of the energy contained in these particles is difficult to imagine. A single proton of the cosmic radiation with an energy of 10^{20} eV contains about 100 billion billion times more energy than a photon of visible light; it has as much destructive power as a weight of 1 kilogram falling to the ground from a height of 1.6 meters. (Note that the rest mass of a single proton is to a kilogram as a kilogram is to the total mass of hundred Earths!) It is obvious that extreme conditions are necessary to impart such tremendous energy to such tiny elementary particles.

Unfortunately, the examination of cosmic radiation itself did not provide any direct clues as to its origin. Since these electrically charged particles are diverted and redirected on their path through the Milky Way by magnetic fields, any directional information is lost, and they hit the atmosphere from all directions.

Pulsars – Energy Beacons in the Universe

The discovery of pulsars in 1967/68 finally shed some light onto the mysterious origins of cosmic rays. The first of these apparently new sources of radiation was noticed by a talented young British scientist. Carrying out observations with the new radio telescope of the University of Cambridge in the fall of 1967, Jocelyn Bell found strange pulses at regular intervals, like some sort of interstellar Morse code. Every 1.337 seconds a peak emerged from the background noise of the instrument. Since the mysterious source did not change its position in the sky, it could not be a distant space probe. The pulses remained even after all possible terrestrial interference had been eliminated, leading a few insiders to joke that they came from the interstellar channel of a galactic radio station. When two more sources were discovered soon after, however, the British scientists did not want to conceal their discovery any longer. In February 1968, they published an

The high time resolution of the ROSAT detectors made possible the distinction of "pulse on" (right) and "pulse off" (left) in the Crab Nebula; this makes it clear that the pulsar itself is much brighter than the nebula.

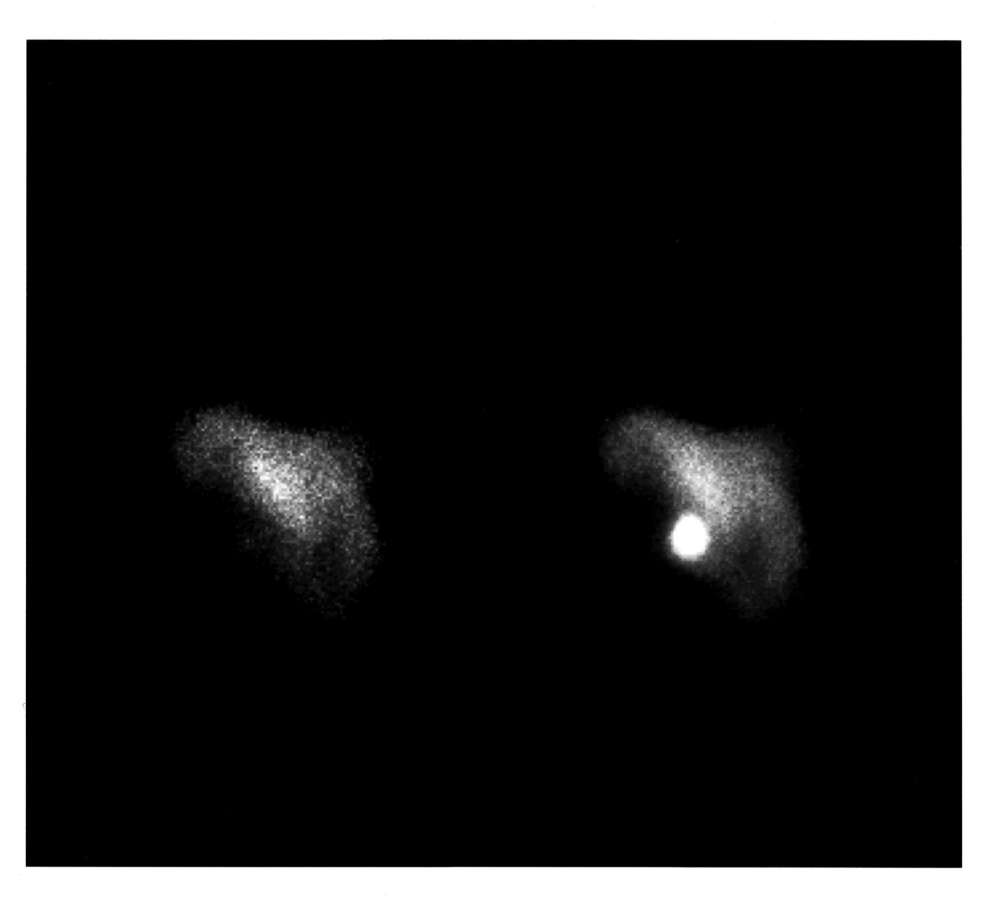

article about "pulsating" radio sources in the journal *Nature*.

Thomas Gold, an Austrian-born astrophysicist at Cornell University in Ithaca, New York, soon provided a plausible explanation for the mysterious objects. The pulses came from rapidly rotating neutron stars, the extremely compact remnants of enormous supernova explosions. The existence of these objects had been predicted in the 1930s. Such stellar corpses, with diameters of little more than ten miles, could rotate in less than a second without being torn apart by centrifugal forces, and they were also supposed to have extremely strong magnetic fields – just the right conditions to accelerate electrically charged particles to very high energy levels.

The discovery of pulsars, and of the pulsar in the Crab Nebula in particular, was a key event for Joachim Trümper, since the Crab Nebula had been known for quite some time as a strong source of synchrotron radiation covering the spectrum from the radio and optical range to the domain of x- and gamma rays. Electrons have to possess energies of a million GeV and above to generate synchrotron radiation over the entire spectrum. At the same time, the energy loss from synchrotron radiation is so large that electrons lose their energy within a few weeks; therefore, they could not be direct remnants of the supernova explosion of 1054 that generated the Crab pulsar.

The Crab pulsar had to be the long-sought source of these high-energy particles. Additional proof came from the observation of a slow increase in the pulsar's rotational period. The neutron star continuously loses rotational energy, and this energy loss is only about ten times larger than the energy contained in the synchrotron radiation of the Crab Nebula. At least ten percent of the rotational energy is transferred into high-energy electrons; the rest probably goes into protons of even higher energy, but without any detectable effects on the Crab Nebula.

Like many of his colleagues who had caught the pulsar fever, Trümper first pondered the theoretical questions around the working mechanisms of such a cosmic particle accelerator. While no definitive answers were available then – and are not today – there were qualitative insights that could be checked by experiments. One was the hypothesis that the ability of a pulsar to accelerate particles had to decline with age, by perhaps a few tenths of a percent per year.

A loss in the central pulsar's accelerating power should be evident in the x-ray brightness of the Crab Nebula: As the energy of the accelerated electrons becomes less, the x-ray signal from the nebula should decrease.

The HEXE Balloon Program

In order to check this prediction, Trümper applied in 1971 to the German Science Foundation to obtain funds for an x-ray balloon experiment. Since a single goal could not justify such an ambitious project, he proposed as well to observe several other variable x-ray sources in the high-energy range (between twenty and one hundred keV). These sources had just been discovered by Uhuru, but the satellite could observe them only at lower energies, from two to six keV.

Trümper, now a professor in the astronomy department at Tübingen University, was given

From 1973 on, several balloon flights to observe cosmic x-ray sources were launched, first from the Institute for Astronomy of Tübingen University and later in cooperation with the Max Planck Institute for Extraterrestrial Physics in Garching, near Munich.

his grant, and he began to develop the balloon-borne High Energy X-ray Experiment (HEXE) with Rüdiger Staubert and other collaborators.

In 1975, Trümper accepted the position of director of the Max Planck Institute for Extraterrestrial Physics in Garching, near Munich. Here the instruments for HEXE could be significantly enlarged and improved. In collaboration with Staubert's group in Tübingen, the program was continued with a total of eight flights, launched in the United States, South America, and Australia.

This balloon program turned out to be the foundation for the further development of x-ray astronomy in Germany. The expeditions, generally lasting several weeks, were impressive experiences for all involved, particularly students. The flights did not always go as planned. It was sometimes necessary to wait days or weeks for the calm winds necessary for launching the huge balloon. Once, when the parachute did not detach from the gondola upon landing and was caught by the wind, HEXE was dragged through the Texas desert for several miles, after which it cut through two high-voltage power lines and destroyed a farm fence before getting stuck in a bush. The expensive equipment, though battered, survived.

These HEXE flights first targeted the Crab Nebula but later concentrated more and more on the bright sources detected by Uhuru, where many questions remained to be investigated in the hard x-ray band.

The Search for the Black Hole

One example for HEXE's contributions was the successful measurement of the magnetic field

The balloon gondola with the HEXE
payload after a hard landing; the
grate-like collimators to improve
spatial resolution are easily visible.

strength of the neutron star Hercules X-1 (see p. 31). Even more bizarre than neutron stars are black holes, whose existence is predicted by Einstein's general theory of relativity. Among many other things, the theory postulates that light is subject to the influence of gravitation. Matter in a black hole is packed so densely that the gravitational forces at its "surface" prevent light and other electromagnetic radiation from escaping.

After neutron stars had been discovered, in the form of radio sources and x-ray binaries, scientists became hopeful of finding at least some clues to the existence of black holes. If such an object were a partner in a binary system, and if matter were attracted from the other companion, then this matter would be more and more heated by friction until finally – in a last outcry before plunging into the back hole – it would emit strong x-rays.

Cygnus X-1 was then considered the best candidate for a black hole; Uhuru had already identified it as a complex, variable x-ray source. Using their balloon-borne HEXE, the scientists in Garching determined that the temperature of the x-ray emitting gas was about 170 million K. Such a high temperature could hardly be explained other than as the heating of matter in its death spiral into a black hole.

Contacts with Moscow

In discussing their observations of Hercules X-1 which had led to the determination of the magnetic field of the neutron star, the group in Garching made contact with Rashid Sunyaev, from the Moscow Institute for Space Research.

Since 1987 the Russian space station Mir contains the research module Kvant at the base of the T-shaped structure (top right, with docked Progress transporter).

But German–Soviet cooperation was a delicate matter. Since Moscow always demanded that agreements with the Federal Republic of Germany include a passage that contradicted the Western interpretation of the status of West Berlin, the German federal government did not officially approve the project and did not provide any money from the federal budget. The scientists were limited to the support available from the Max Planck Society. This was the only way to get HEXE onto the space station. In 1982, an agreement was finally signed in Moscow, initiating a cooperation that anticipated, on a small scale, what would later become known as perestroika and glasnost. Five years later, the experiment from Garching flew to the Mir space station and was installed on the Kvant expansion module, in which British and Dutch research groups also participated with their own instruments.

The first highlight of Mir–HEXE was the detection of hard x-rays from the famous Supernova 1987A in August 1987. This radiation originates from the decay of radioactive elements generated in fusion reactions during the explosion. Many more measurements followed, and even a decade after its launch in March 1987, Mir–HEXE is still active.

By 1987 the x-ray satellite ROSAT should already have been operational. But its development was both slow and convoluted.

As a theoretical physicist, Sunyaev had studied the structure of matter under extremely strong magnetic fields and had predicted the existence of cyclotron resonance lines in the x-ray domain. From this first exchange of ideas across the Iron Curtain, a fruitful cooperation developed, which finally led to a joint project. In spite of its successes, the HEXE balloon program was hampered by the limited flight time of the balloon. To overcome this handicap, Trümper proposed to Sunyaev that HEXE be deployed on the Soviet Mir space station, whose launch was then planned for the mid-eighties.

This possibility was particularly attractive because the Germans' cost would not greatly exceed that of the balloon program. The project was meant to be carried out without the extensive documentation required by NASA, which demanded the space qualification of every last nut and bolt.

ROSAT – The ROentgen SATellite Project

The first step toward ROSAT was a proposal Joachim Trümper and his group in Garching made in 1975 to the Federal Minister for Research for a

During the preparations for the development of ROSAT, this Wolter telescope, with an aperture of 32 centimeters (12.5 inches), was manufactured and tested in several rocket flights.

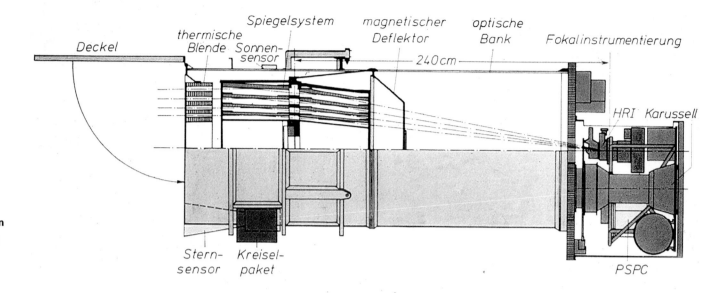

Deckel · thermische Blende · Sonnensensor · Spiegelsystem · magnetischer Deflektor · optische Bank · Fokalinstrumentierung · 240 cm · HRI · Karussell · Stern-sensor · Kreisel-paket · PSPC

The twofold grazing-incidence reflection on a combination of a parabolic and a hyperbolic mirror makes it possible to obtain images of x-ray-emitting objects; the effective mirror surface – and therefore the x-ray yield of the telescope – is increased by concentrically stacking several mirror shells.

Translation of German terms
   (English in parentheses):
Deckel   (cover)
thermische Blende   (thermal
   screen)
Sonnensensor   (sun sensor)
Spiegelsystem   (mirror system)
magnetischer Deflektor
   (magnetic deflector)
optische Bank   (optical bench)
Fokalinstrumentierung   (focal
   plane instruments)
Karussell   (carousel)
Sternsensor   (star sensor)
Kreiselpaket   (gyroscope
   assembly)

satellite with an x-ray telescope. The idea received early encouragement from the Maier-Leibnitz-Commission, tasked to evaluate plans for large scientific projects for the ministry. The commission chose the x-ray satellite for development along with two other projects, and the European Science Foundation and U.S. Academy of Sciences expressed strong interest in the concept as well. Preliminary technical studies were invited and carried out for both the satellite and its payload, an imaging x-ray telescope.

However, there were serious doubts about the costs. The Federal Republic of Germany had already completed a number of national satellite projects, including two solar probes, Helios 1 and 2. These projects had led to a rule of thumb that each kilogram of payload in orbit would cost about one million German marks. ROSAT, with its projected weight of about one thousand kilograms, worked out to the staggering sum of one billion marks. But the scientists countered that most of ROSAT's mass

was contained in the telescope, which would be much cheaper than the dense packages of detectors and electronics in earlier payloads. They began to call their project ROBISAT (ROentgen BIllig-SATellit, or Cheap Roentgen Satellite).

In the end, ROSAT's launch mass came to almost two-and-a-half metric tons, inflated in part by the politically motivated demand to make it an international project. The cost for ROSAT at launch of 260 million German marks, as given by the Ministry for Research, corresponds to about 100,000 marks per kilogram – lower than the first estimates by a factor of ten. ROSAT remained a comparatively cheap satellite, although the name ROBISAT was changed to ROSAT for aesthetic reasons at the beginning of the eighties.

Grazing Reflections

The basic design of imaging x-ray optics was developed during the early nineteen fifties by the physicist Hans Wolter, who was then at Kiel University developing an imaging x-ray microscope. It had been known since the thirties that x-rays could be redirected and focused into an image using reflections on a surface at very shallow angles. In this way, nature can be outwitted: Because of their short wavelengths, when they strike at steep angles, x-rays pass between the atoms of a surface and then are quickly absorbed. But at very shallow angles, the gaps between atoms of the mirror surface appear much smaller, so that the x-rays are bounced back. There was only one problem: a simple parabolic focusing mirror, like those used in optical telescopes, generated only highly distorted images for x-rays at grazing incidence.

Wolter found a way around this problem. After the grazing reflection on a parabolic ring, he directed the x-rays onto a ring of hyperbolic shape and thus obtained a useful and sharp image. The longitudinal section of this combination of mirrors resembles a long, conical tube with a slight kink in the middle. The mirror surface has to be extremely smooth, so that the x-rays can be properly reflected without being scattered. Everybody who has looked at a surface under a shallow angle against the light knows how prominently even small irregularities or dust particles stand out. This situation is exacerbated in the x-ray domain, where wavelengths are only one-thousandth those of visible light.

Technical Preparations

To avoid such scattering effects, the so-called micro-roughness of the mirror surface had to be pushed back to almost atomic dimensions. It could not exceed a few atom diameters (or several ten-millionths of a millimeter). It was a great technological challenge to manufacture such mirrors.

By 1973, the decision had been made in Tübingen to begin developing x-ray telescopes. But it was not easy to convince a German optical company to undertake such a project. Aside from the required high precision and low micro-roughness of the mirrors, the unusual tube geometry was a major deterrent.

Joachim Trümper and Heinrich Bräuninger therefore counted themselves lucky that Horst Köhler, from Carl Zeiss in Oberkochen, agreed to collaborate with them in creating this new technology. The project was planned in four steps. The first was to make tests on flat mirrors to identify the optimal methods for polishing.

The scientists' material of choice was zerodur, a ceramic material developed by the Schott glass works in Mainz that exhibits negligible thermal expansion. It is used for manufacturing stove tops as well as in optical telescopes – two applications where expansion and changes in shape as a result of temperature variations are undesirable. American scientists in particular doubted that a micro-roughness of only a few ten-millionths of a millimeter could be achieved, because of the crystalline structure of the ceramic material. Polishing tests and measurements of the resulting surfaces refuted those doubts.

In parallel with the tests of materials, a few dozen parabolic metal mirrors were manufactured and used in two sounding rocket experiments during the late seventies. This was the second step of the program.

As a third step, three small Wolter telescopes

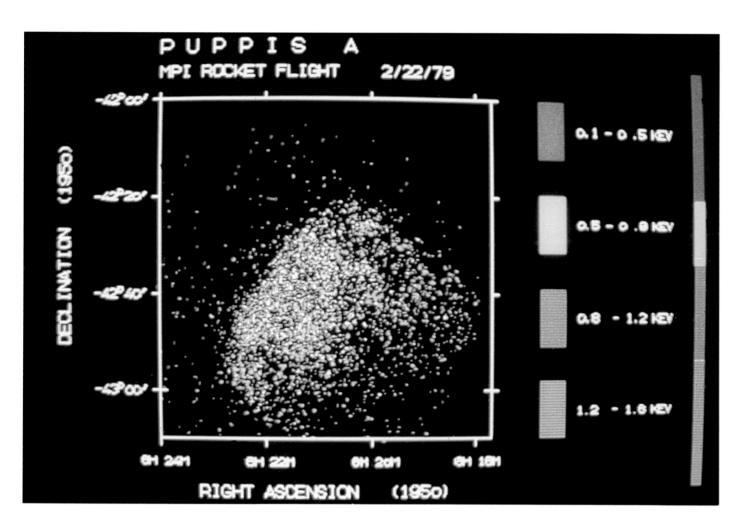

The first "European" x-ray image, and also the first x-ray color image of an extended source, showing the supernova remnant Puppis A; the image was obtained in 1979 during a rocket flight with the 32-centimeter Wolter telescope and a precursor of the ROSAT PSPC of the Max Planck Institute for Extraterrestrial Physics (see also page 15).

of 32 centimeters diameter each, were built by Carl Zeiss; together with imaging x-ray detectors, developed and built by Elmar Pfeffermann and Horst Hippmann at MPE, they became true precursors of ROSAT on the occasion of three sounding rocket experiments in 1979, 1982, and 1987. The memorable flight on February 22, 1979, launched from Woomera in Australia and dedicated to the exploration of the supernova remnant Puppis A, yielded two firsts: It generated the first x-ray image by a non-U.S. experiment and the first x-ray "color" image in history. These rocket flights were geared mainly toward tests of the technology and demonstrations of feasibility. The scientific return, as important as it may have been, was limited by

the flight time of only five minutes.

The scientists invested great effort in the calibration of telescopes and imaging detectors. Under the direction of Heinrich Bräuninger, increasingly powerful test stands were built, beginning with a 6-meter setup in Tübingen and extending to the 130-meter test facility built at MPE for ROSAT in the early eighties and in continuous use for various projects since then.

Difficult Production

The starting point for the production of the mirrors was several large, solid blocks of zerodur. Out of these, eight single cylinders were milled and their

The x-ray optics of ROSAT are included in the *Guinness Book of Records* as the "smoothest mirrors in the world."

interior and exterior surfaces smoothed. Grinding, polishing, and assembly were done by Carl Zeiss in Oberkochen. Since the (inner) mirror surfaces deviate significantly from a simple spherical shape, they posed high demands on the grinding and polishing operations. The tools had to follow two different but extremely precise curvatures with no deviation. In addition, the parabolic and hyperbolic surfaces of the mirrors had to have exactly circular cross-sections. In order to achieve an undistorted image, the cross-sections of the mirrors could not deviate from exact circles by more than a micron (a thousandth of a millimeter). The longitudinal sections of the mirror segments had to conform to the prescribed parabolas and hyperbolas with similar precision.

To avoid breakage, the delicate mirror seg-ments, only a few centimeters thick, had to be very carefully mounted in special support structures for grinding and polishing. Special emphasis had to be given to stress-free mounting: If the outside of the mirror cross-section were forced into the shape of a slight oval, followed by the required precisely circular grinding of the inside, this circular mir-ror surface would have relaxed into an oval upon removal from the support structure. Fortunately, the manufacturing problems with the mirror of the Hubble Space Telescope had not yet become pub-lic; if they had, the pressure on the engineers at Carl Zeiss would have been even greater than it was.

Polishing each single mirror took about six to eight weeks, including the frequent intermediate tests. A testing device, specially developed for

The break-proof and vibration-resistant assembly of the ROSAT mirrors posed a major technical challenge.

telescope, with a diameter of 83 centimeters, achieved an angular resolution of 5 arc seconds, surpassing all previous instruments of this kind. But the road to this achievement was difficult and full of technical challenges.

One difficulty had to do with depositing a thin layer of gold onto the mirrors to increase their reflectivity; this behavior is influenced by the density of the material and particularly by the so-called quantum number of the element. The higher the quantum number, the larger the number of electrons in the vicinity of the nucleus on which x-rays can be reflected. All attempts to evaporate in vacuum a 100-nanometer layer of gold onto the mirror showed unsatisfactory results. This changed only after a different method was tried, in which gold atoms are sprayed, or "shot," onto the surface. This worked so well that the ROSAT mirrors even made it into the *Guinness Book of Records*. If the mirrors were enlarged to the size of Lake Ontario, their irregularities would correspond to waves one-twentieth of a millimeter high.

How to Glue Glass to Metal?

A special challenge arose when the eight ceramic mirrors had to be assembled one inside the other to form the telescope. The individual "tubes" had to be precisely adjusted relative to each other and then solidly affixed, and in the case of glass this is possible only by stable gluing. But onto what can one glue glass of extremely low thermal expansion? In a satellite, temperature changes are unavoidable, and these would lead to stresses

ROSAT, revealed deviations down to a hundred-thousandth of a millimeter from the prescribed shape. In the end, all departures of more than a ten-thousandth of a millimeter had been removed, and between any remaining "peaks" of more than 0.3 nanometers there had to be a distance of at least 3 millimeters.

This precision is the reason that the ROSAT

between the mirrors and the support structure, deforming the mirrors.

After long discussions, an alloy of iron, nickel, and steel, called super invar-steel, emerged as a solution, as it had low thermal expansion. To further protect the mirrors against thermal expansion, they should be attached to the support structure by thin, L-shaped sheet metal. But could these thin pieces of sheet metal hold the mirrors in place through the strong vibrations of the launch, or would ROSAT's mirror break away and enter its own orbit?

Computer simulations by aerospace engineers to predict the behavior of the mirrors and the sheet-metal flanges gave conflicting results, so that the telescope manufacturer undertook yet more calculations. These results supported the "optimists,"–and indeed, after a structural model was built and tested for the expected level of vibration, the entire assembly held together.

ROSAT Goes International

By this time, the project had evolved into an international collaboration. During the late seventies an opportunity for add-on experiments had been announced within the European Space Agency (ESA), resulting in British participation. Astronomers of the University of Leicester wanted to contribute a wide-angle camera for the extremely far ultraviolet, to extend the spectral range of ROSAT into longer wavelengths. At the beginning of the eighties, negotiations with NASA led to a second extension. The Americans agreed to provide a high-resolution imaging detector (HRI), an improved version of the equipment flown on

Frontal view of the ROSAT mirror system; the circular grille stabilizes the four concentric mirror shells.

the Einstein satellite, in order to continue these measurements with a more capable telescope. In exchange for use of the telescope and satellite, NASA was prepared to carry ROSAT into orbit aboard the space shuttle.

In the meantime, intense work on the German ROSAT detector continued at MPE in Garching under the direction of Elmar Pfeffermann. This detector was intended to register spatial, energetic, and temporal information for each arriving x-ray photon and to surpass the sensitivity and resolution of all previous instruments as far as possible. It would have a gas proportional counter with three planes of wire screens. On the two outer ones negative high voltage was applied, forming the cathodes, while the central one was to be the anode, with corresponding positive high voltage. The x-rays focused by the Wolter telescope are absorbed by the gas in the detector–a mixture of argon, xenon, and methane–and they produce tiny discharges between the wire planes, which are then registered by the closest wires. A mi-

The Position-Sensitive Proportional Counter (PSPC), the x-ray color camera developed at the Max Planck Institute for Extraterrestrial Physics, with its large four-part filter wheel; the central entrance aperture and the boron filter on the right are visible.

croprocessor determines the exact location of the "impact" from the measured pulses. To achieve the desired high spatial resolution, the wires have to be positioned with an accuracy of only a few microns, which could only be achieved using very precise wiring specially designed for the task at MPE.

Imaging of X-Rays

The energy resolution properties of the proportional counters can divide the range from 0.1 to 2.4 keV into four bands, corresponding in principle to colors in the optical range – ROSAT can therefore produce "color" images of the x-ray sky. These "colors" provide much physical information. Generally, the less difference in intensity between the short-wave "harder" x-ray range and the long-wave "softer" band, the hotter the source. Optical astronomers obtain a first estimate of the temperature of a star in similar fashion: Cool stars are

brighter in the long-wave (red) part of the visible spectrum than in short-wave (blue) light, and therefore appear reddish, while hot stars are white or bluish because of their higher intensity at short wavelengths.

To avoid counting particles of the cosmic radiation, both proportional counters aboard ROSAT are surrounded by "veto-counters" on five sides; if any one of them registers an event at the same time as the actual detector, the acquisition electronics "knows" that this hit was caused by a charged particle, not an x-ray photon. These interfering signals are automatically discarded, so that more than 99% of the cosmic background signal can be suppressed.

The entrance window of the detector has to satisfy extreme demands as well. It has to be thin enough to let x-rays pass through, but at the same time it has to form an efficient barrier between the gas inside the detector (at one-and-a-half times the atmospheric pressure) and the vacuum of space outside. In the end, a polypropylene membrane with a thickness of only one thousandth of a millimeter was selected; a mesh of thin metal wires (with a thickness of 0.025 and 0.2 millimeters) and a radial support stiffen this membrane of more than 50 square centimeters. An electrically conducting layer of graphite prevents electrostatic charging of the membrane, and an additional outer coating prevents ultraviolet light from entering the detector.

In spite of these sophisticated precautions, the detectors do not yield a uniform image when the entrance aperture is illuminated with a parallel beam of x-rays; the uneven intensity distribution results from, among other things, the arching of the windows under the large pressure difference. To correct for such distortions, images of various calibration masks were taken at the x-ray test facility at the MPE, resulting in large correction tables, which allow the true positions of the x-ray photons to be reconstructed, resulting in an ideal, distortion-free image.

Filter wheels with four sectors are located in front of each proportional counter. During normal measurements, an open aperture is used, allowing the photons focused by the telescope to pass into the detector. For determining the instrumental background noise and the interfering cosmic radiation, a closed position prevents x-rays from entering. The third sector contains a boron filter for splitting the long-wave part into two additional colors, and the fourth sector holds three radioactive substances for the energy calibration of the detectors.

An Artificial "Optic Nerve"

The "brain" of the focal plane instruments consists of complex electronics containing two microprocessors among other elements. It was developed at MPE by Horst Hippmann, and includes more than three hundred circuit boards as well as over six kilometers of cable connections. For safety, all electronic subsystems are duplicated, so that a failure can be immediately corrected. While a third of the electronic systems deal with general health and safety support tasks, much like the human brain stem, the remaining two thirds are responsible for the satellite's "sensory organs." For instance, the high voltage for the detectors has to be generated and kept constant, and the electrical

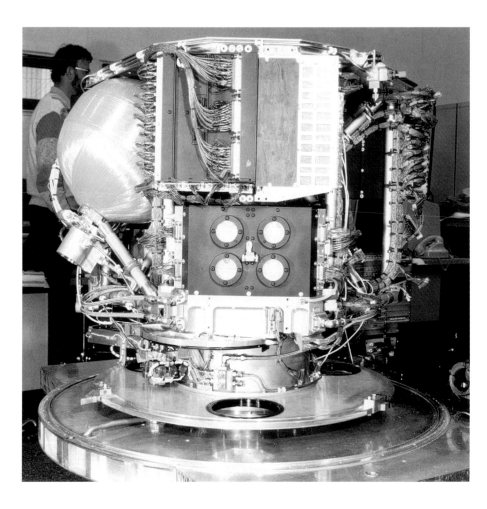

impulses of the detectors have to be amplified, analyzed, and digitized, so that they can be converted by the microprocessors into image information. Finally, the central processing unit of the satellite has to receive and store commands from the ground station and control the remaining systems when there is no contact with the ground. Should any of the subsystems report a malfunction, the central processing unit has to bring that subsystem into a save state to avoid further damage. This mechanism also has to work for problems reported by the on-board computer, such as loss of attitude information.

A total of twenty thousand lines of computer code had to be written and painstakingly tested. In principle, it covered all anticipated requirements for the operation of the satellite, but the flexible

The complete ROSAT focal plane instruments before assembly into the satellite; the proportional counter (front) is in the observing position, defined by the aperture in the bottom plate.

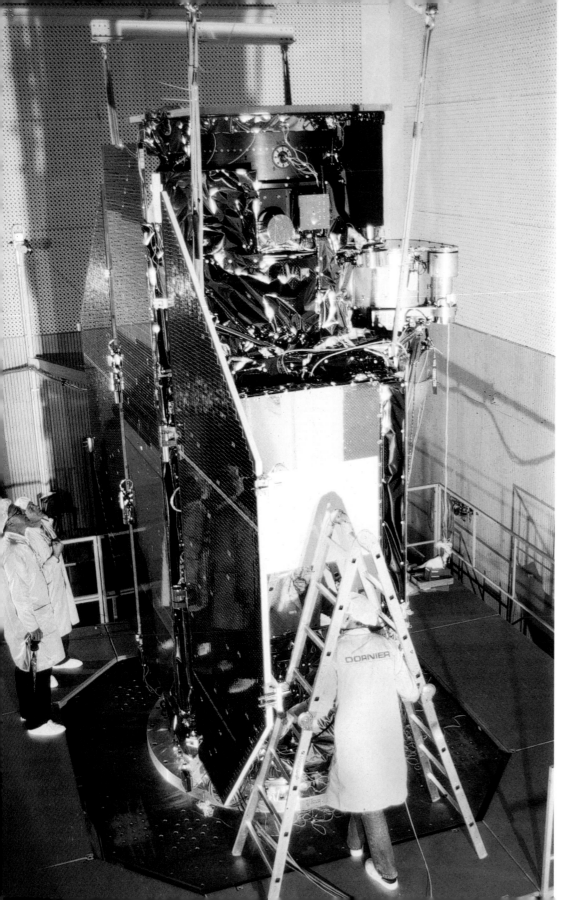

ROSAT is the largest and heaviest national European research satellite.

software structure allowed for later modifications from the ground, so that additions could be made to the programs or entire parts exchanged—a feature that has proven very useful.

Building a Satellite

The satellite carrying the scientific payload was built by the aerospace company Dornier Systems in Friedrichshafen, Germany. Messerschmidt-Bölkow-Blohm delivered the attitude determination and control system, among other parts. This system consists of two electronic star trackers (CCD cameras), which are directly attached to the x-ray telescope and achieve an angular resolution of one arc second. While the telescope carries out measurements, these cameras are read by the on-board computer once per second, and the positions of the stars in their field of view are compared to the expected positions.

Potential deviations are relayed to the navigational gyroscope subsystem, which has to assure the stable pointing of the satellite in the correct direction to permit long-term measurements and high spatial resolution. Its three gyroscopes are mounted at right angles to each other, similar to our three-dimensional concept of length, width, and height; according to their role, they are called simply X-, Y-, and Z-gyro. A fourth gyroscope serves as a backup.

If the satellite is made to drift from its expected orientation by exterior forces such as friction in the atmosphere, interactions with Earth's magnetic field, radiation pressure from the sun, or irregularities in Earth's gravitational field, the star tracker cameras register this deviation. The on-

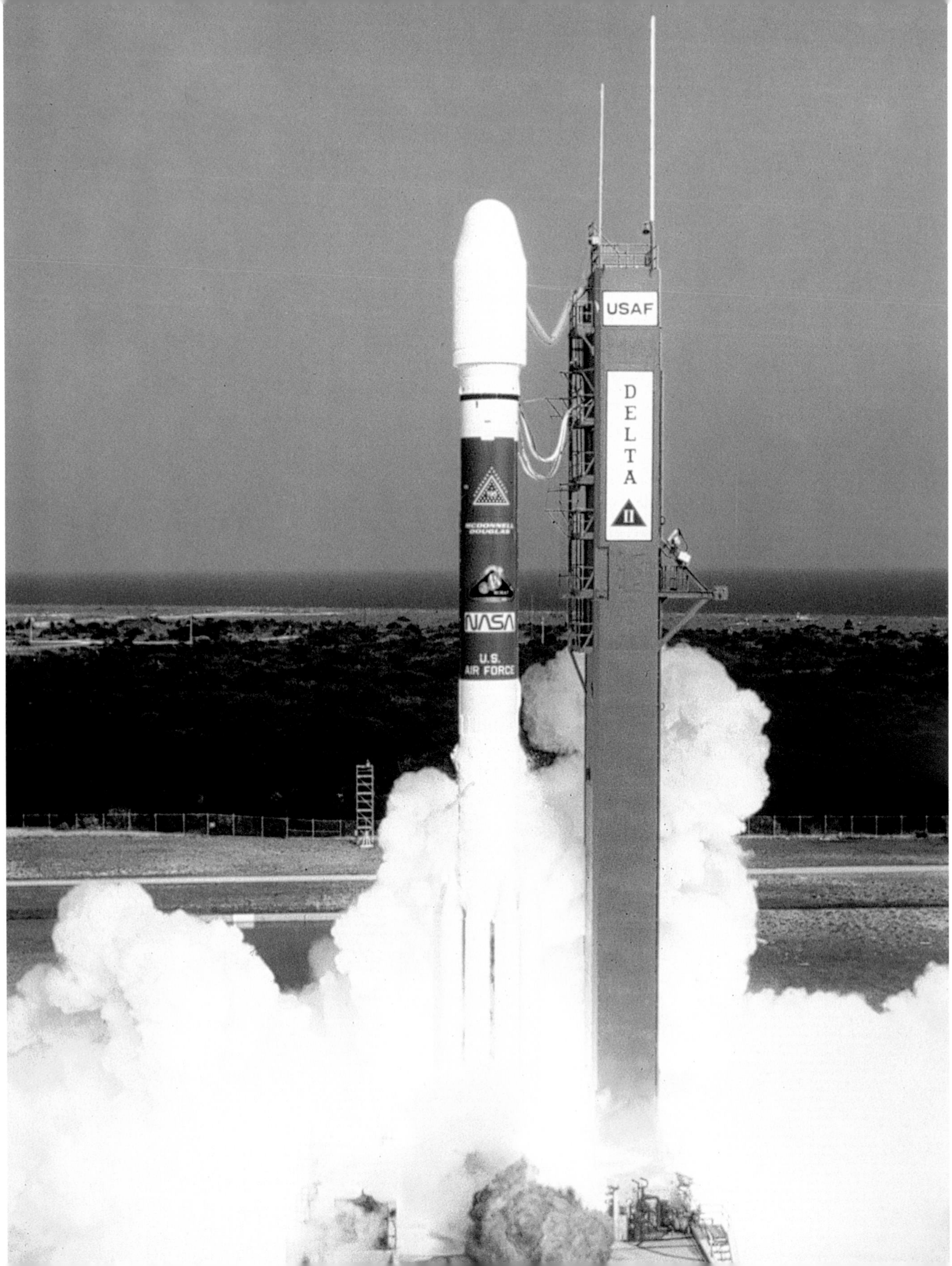
USAF
DELTA
II
NASA
U.S.
AIR FORCE

On June 1, 1990, at 5:48 P.M. local time, ROSAT was carried into space from Cape Canaveral atop a NASA Delta-II rocket.

board computer then activates reaction wheels, which counter the disturbing rotation. A "jitter" of about ten seconds of arc remains; though small, this is still large compared to the intended spatial resolution of one second. This goal can only be achieved on the ground by correcting the x-ray image data with the information from the star tracker cameras, thereby deblurring the images.

To avoid a gradual "overload" of the reaction wheels, the attitude control system also contains three magnetic torquers at right angles to each other; they can generate a magnetic field in any direction necessary. In this way, an appropriate counterrotation of the satellite can be induced with respect to Earth's magnetic field, compensated by the required discharge of momentum in the reaction wheels.

The end result was a satellite of 2.4 metric tons, meant to be carried into orbit by the space shuttle in 1987. However, all NASA launch plans fell into disarray after the explosion of the space shuttle *Challenger* on January 28, 1986. In negotiations for a new launch date for ROSAT, NASA could only offer a shuttle flight "after 1993." The ROSAT engineers, however, considered this a dangerously long time for the satellite to remain in storage. But thanks to the support of NASA and of all German organizations involved in the project, a solution was found that did not require major changes in the satellite: Instead of the space shuttle, ROSAT would now be launched by one of the dependable and proven Delta II rockets.

The long wait finally came to an end on June 1, 1990: Many project scientists, engineers, and technicians, as well as representatives of the German Federal Ministry for Research and the German Space Agency, watched as the rocket carried ROSAT into the blue sky over Cape Canaveral at 5:48 p.m. local time and disappeared within minutes in a northeasterly direction. This was the beginning of the most successful satellite mission ever to explore the invisible sky.

Commands from Bavaria

Eighteen minutes after launch, while ROSAT was still connected to the last stage of the rocket, the German Space Operations Center (GSOC) in Oberpfaffenhofen, near Munich, made its first contact with the satellite; its orbiting altitude of 580 kilometers had been chosen so that ROSAT would cross the reception area of the ground station five to six times a day for about ten minutes each time. Since Earth rotates under the satellite orbit, these contact times are bunched into about eight hours, followed by sixteen hours of no communication between satellite and ground station.

This approach, which eliminated the need for a second ground station, still provided enough time to receive scientific data and status information from the satellite and to transmit commands for the next set of measurements.

In the first days in orbit, a comprehensive test and check program was conducted from Oberpfaffenhofen, transmitting more than twenty thousand commands to ROSAT via the 15-meter antenna in Weilheim. First, the 12-square-meter solar panels were unfolded and the antenna deployed. Then the cover of the x-ray telescope, which protects the instruments from direct sunlight, was opened. One by one, the various subsystems followed: power

View of the control center of the German Research Institute for Aeronautics and Space Flight (DLR) in Oberpfaffenhofen near Munich: Here, the health and safety of the satellite are monitored, the engineering and scientific data arrive, and commands to the satellite are generated.

This fifteen-meter (fifty-foot) antenna in Weilheim, near Munich, provides communication between ROSAT and the German Space Operations Center in Oberpfaffenhofen five to six times per day.

supplies, heating and cooling systems, and the data acquisition and analysis system, including the transmission subsystem and the command storage area. In addition, the recall of data and their rapid processing on the ground for technical and scientific purposes were tested.

The highlight of the preparatory phase arrived with the tests of the ROSAT telescope and the British wide-angle camera. After the power supplies for the various instruments in the focal plane were switched on, the carousel with the three imaging x-ray detectors was rotated until the position-sensitive proportional counter (PSPC) was at the focus of the telescope. Then the high voltage for the PSPC was switched on, and the camera's sensitivity was measured using built-in slightly radioactive sources. In the meantime, a particle counter measured the background radiation from the Van Allen belts. These belts contain energetic, electrically charged particles, which are trapped in the terrestrial magnetic field and oscillate between its poles, giving off energy.

With one small exception, no problems were uncovered in the operation of the satellite. The test phase took only a little more than two weeks, and then the satellite transmitted its first scientific data. A few days later, a six-week scientific verification phase began, involving about two hundred single observations. Even in this early phase of the ROSAT mission, the images were of unsurpassed quality.

On August 1, two months after launch, the first complete color survey of the x-ray sky with an imaging x-ray telescope began. During the following six months, ROSAT and its imaging x-ray detectors scanned a narrow strip of sky with

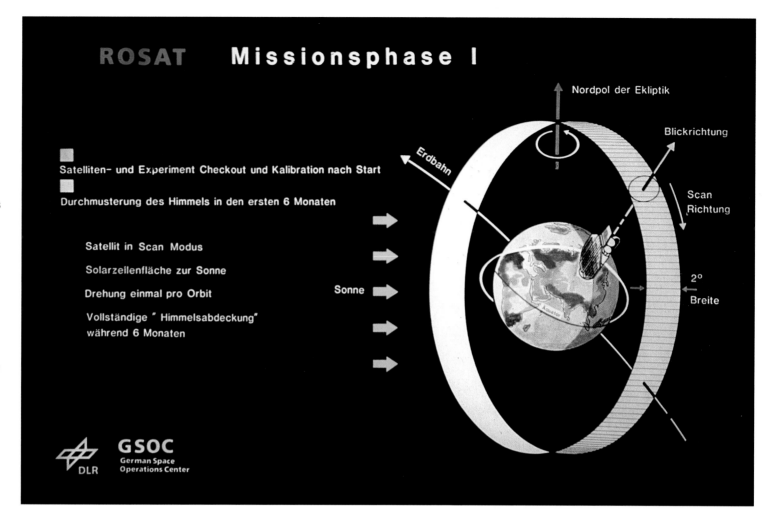

every orbit around Earth; the width of this strip was determined by the diameter of the field of view of the telescope. The gradual movement of the Earth around the sun rotates this strip, so that it covers the entire sky in half a year.

Early Morning Shock

Shortly before the end of the survey, ROSAT almost became the victim of an accident, which is still not completely understood: On January 25, 1991, during one of the usual periods without communication with the ground, the satellite suffered a complete failure of its attitude control computer and began to drift. As a result, the solar panels did not capture enough light from the sun and could not provide enough energy for the satellite. In this case, the central computer of the satellite was supposed to switch control automatically to the available backup unit of the attitude control system, but somehow this function was blocked as well. Even the basic safe mode, specifically designed for such cases, failed. This would have positioned the satellite so that the solar panels remained optimally illuminated, guaranteeing the supply of energy.

As the satellite checked in again at Oberpfaffenhofen "with a soft voice" in the early morning hours of January 26, the satellite operator on duty suffered quite a shock: The data on the status of

ROSAT made it clear that a complete power failure of the satellite was imminent, that all experiments had switched off, and that the attitude control system was providing hardly any data. Data transmission itself was affected as well. Since after only nine minutes ROSAT would be inaccessible again for another hour and a half, there was not enough time for an immediate rescue attempt.

By the next passage, shortly after 5 o'clock, system engineers had been assembled at the control center at GSOC to analyze the situation. After eliminating data transmission problems on the ground, it became clear from the information from ROSAT that the satellite was in serious danger.

Since the solar panels were not supplying ROSAT with energy, the charge of the batteries had fallen below a lower limit, and all nonessential systems on board had been switched off. It became painfully obvious to everybody – even without consulting the extensive documentation about the satellite – that the situation was close to the worst accident scenario imaginable and could easily end the mission.

During the night, the staff in Oberpfaffenhofen contacted the engineers at the company in Friedrichshafen that had built the satellite. However, at the next passage, shortly after seven in the morning, the satellite remained silent, and all attempts to reestablish the communication link failed.

A race against time ensued. With the greatest possible speed, procedures to rectify the malfunctions were developed. The corresponding command loads for the satellite were prepared, and NASA was contacted to obtain further op-portunities for contact with ROSAT via additional ground stations.

When ROSAT came within reach of the Weilheim ground station for the fourth time that day, at about 8:30 A.M., the control center was able to switch on the carrier signal aboard the satellite, but normal data transmission from the satellite to the ground did not restart. In parallel, a set of commands was transmitted to ROSAT, attempting to activate the backup attitude control computer and to stabilize the satellite. Then contact was again lost.

During the remaining hour and a half until the fifth and last contact on that Saturday with the Weilheim antenna, a program to reinitialize the central on-board computer was developed and then transmitted to the satellite at 10:12 A.M. Indeed, telemetry data began to appear again on the screens of the control center; they showed that the satellite had "caught" itself and was now in a safe attitude, with the solar panels charging the batteries. ROSAT had been saved.

Six more times that day, the Oberpfaffenhofen control center made contact via NASA ground stations with their weak patient. Since the collapse of the electrical system had left the entire configuration of the satellite in disarray, every subsystem had to be restored in sequence. Then it became apparent that parts of the scientific payload had been damaged. The satellite had pointed directly toward the sun during its uncontrolled drift, so that the concentrated heat at the focus destroyed the entrance window of one of the proportional counters and one of the filters of the British wide-angle camera. Luckily, a second identical proportional counter had been installed as a backup, so the loss

This x-ray map of the sky in galactic coordinates contains almost 80,000 single sources. The map was constructed from data of the ROSAT sky survey. The color scale from red via yellow, green, and blue to violet indicates the spectral characteristics (from soft to hard x-ray spectra), and the size of the dots corresponds to the total intensity of the radiation from the source.

ROSAT ALL-SKY SURVEY Sources

All-Sky Survey

The soft diffuse x-ray background radiation originates mainly from hot interstellar gas. Its spatial distribution in the sky shows several different components: Soft radiation (at 0.25 keV, red) comes from the local hot bubble, a 100,000 year-old supernova remnant in the solar vicinity, but also from the galactic halo. At higher energies (0.75 keV, yellow, and 1.5 keV, blue) more distant galactic supernova remnants and hot stellar winds dominate. The small regions (white) are either compact supernova remnants or galaxy clusters.

did not diminish the scientific return from the mission. The backup detector operated flawlessly until its gas supply ran out during the summer of 1994.

Very early in the dramatic rescue operation, ROSAT had been placed into a safe mode configuration to minimize further damage; it remained in this state for almost two weeks, until a team of experts from industry, the Max Planck Institute for Extraterrestrial Physics, and the satellite control center had narrowed down the cause of the malfunction. By careful analysis this team eliminated step by step the possibilities of faulty hardware design, faulty components, and software errors. The most probable cause was the transient malfunction of a sensitive component in the interface between the attitude control and the central computers, possibly caused by a short-term increase in cosmic ray intensity.

Solar physicists had registered a strong eruption near the solar surface a few hours before ROSAT's malfunction; the particle shower associated with this event could have reached Earth and the satellite at the time in question. Since such an event could happen again, several improvements in the attitude control computer's software were derived from the anomaly analysis and loaded into the computer after intense testing.

The Most Accurate X-Ray Map

Because of the accident shortly before the end of the survey, it could not be completed until the summer of 1991, when ROSAT could observe the remaining strips of sky without pointing too close to the sun. Several months later, Joachim Trümper presented Federal Minister for Research Heinz

Riesenhuber with the first copy of the complete ROSAT sky map.

Its latest version contains about eighty thousand cosmic x-ray sources, or about fourteen times more than had been known before ROSAT: between twenty-five and thirty thousand stars in the vicinity of the sun, between twenty and thirty-three thousand active galaxies and quasars at large distances, about five to ten thousand galaxy clusters, and finally, more than one hundred supernova remnants, which turned out to be particularly photogenic because of their size and shape.

Compared to the previous sky survey, carried out by the American HEAO-1 satellite and containing only 840 objects, ROSAT was capable of identifying sources one hundred times fainter and determining their positions to an accuracy of about thirty arc seconds – about two to three times more precise than trained observers measuring stellar positions with the naked eye. This improved positional accuracy makes it easier to identify x-ray sources with known optical or radio counterparts, which in turn allows us to obtain supplemental information about the sources (for instance about their distance).

Astro-Navigation for ROSAT

In addition to the grave accident at the end of January 1991, certainly the most critical moment in the success story of this satellite, several smaller malfunctions occurred but were overcome. The failure of several components of the attitude control system must be particularly mentioned.

In November 1990 one of the star tracker cameras had failed. But through software changes

THE FIRST ROSAT SOURCE CATALOGUE Sources

Aitoff Projection
Galactic II Coordinate System

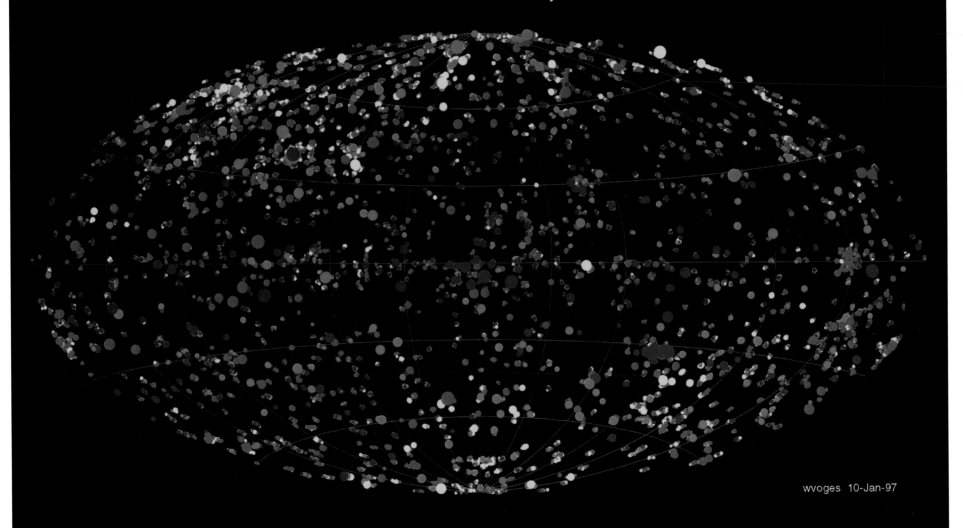

wvoges 10-Jan-97

Energy range: 0.1 - 2.4 keV
Number of PSPC sources: 74310
Hardness ratio: -1.0 I -0.6 I -0.2 I 0.2 I 0.6 I 1.0   (soft -> hard : red - yellow - green - blue - violet)

More than 70,000 objects of mainly extragalactic nature are contained in this first ROSAT map of x-ray sources, obtained during a period of three and a half years after launch in pointed mode with the PSPC x-ray camera. (For color coding, see the caption for the figure on page 61.)

both on the ground and aboard the satellite, the control center was able to achieve the required fine pointing during measurements despite the one-eyed performance.

Two of the four navigational gyroscopes failed over time, and a third delivered data of reduced accuracy.

In the meantime, the ROSAT all-sky survey had been completed, giving way to pointed observations in which a single source would be acquired and monitored for long periods. This mode requires a high degree of pointing accuracy to achieve the intended high spatial resolution for the measurements. Thanks to the gyroscopes, the satellite knows its pointing direction during a slew from one source to the next while it finds its new target in the sky. Upon reaching it, the star tracker camera image is compared to the positions of the stars stored in the computer, resulting in the fine pointing of the telescope. For two months ROSAT had observed about thirty sources per day in this way, until the Y-gyro failed on May 12, 1991. For some time after this, the satellite was able to observe just one source per day, with the pointing of the telescope being corrected "by hand" from the ground station.

However, scientists used this delay in a busy observing schedule for interesting and revealing long-term observations–the first survey of the Andromeda galaxy for faint x-ray sources, for instance, was carried out during that time.

It took five months to develop and test pointing methods that did not use the Y-gyro. This was followed by two years of largely normal observing operations, until the Z-gyro failed on November 17, 1993. This failure again tested the scientists' creativity; in the end, they developed a navigational approach, applying the principles used by seafarers for centuries: the determination of the position of the sun and the direction of the local magnetic field. The sensors for the sun and for measuring the terrestrial magnetic field were originally intended only as safety devices: the sun sensor was only supposed to prevent the telescope from pointing anywhere near the sun and so burning out its instruments. But now the two "sensory organs" were combined with the two remaining gyros to create a new attitude control system, unique in its way.

Since the sun was now needed as a reference point, slews during occultation of the sun by Earth were no longer possible, causing restrictions on the selection and sequence of targets in mission planning, but this did not significantly diminish the continued successful operation of the satellite.

This failure and the way it was corrected turned into a success for the German aerospace industry, which had shown on a large scale that it was possible to overcome hardware malfunctions with intelligent software. ROSAT's novel attitude control concept, born of necessity, directly led to the award of a major satellite construction project against international competition.

X-Ray Astronomy in Our Galaxy

In late March 1996, an unusual celestial phenomenon attracted the attention of many observers, and of the media in particular. A bright comet, the brightest in at least twenty years, had appeared in the night sky. The Japanese amateur astronomer Yuji Hyakutake had discovered it as a faint nebulous spot two months earlier.

When it turned out only a short time later that the new comet, in traditional fashion bearing the name of its discoverer, would pass Earth at a distance of only 15 million kilometers and develop into a potentially record-setting object, the public and professional astronomers alike became interested. The former hoped for spectacular views and photos, while the latter expected revealing data about size and composition of this cosmic traveler.

All available telescopes were trained on C/1996 B2, as comet Hyakutake was called officially. By late February, astronomers at the European Southern Observatory (ESO) in Chile had found strong indications of molecular carbon, cyanide, oxygen, and ionized carbon monoxide in the comet's spectra. American scientists discovered parts of molecular compounds, such as from hydrogen cyanide and water, broken up by UV radiation from the sun. In the meantime, astronomers at the Pic-du-Midi Observatory in the French Pyrenees observed the gas jets on the "morning side" of the comet core, their highest activity at "noon," and their disappearance on the "evening side"; they deduced a rotational period of the comet core of about six hours.

Radio astronomers also measured the signals from the comet, from which they inferred the ratio of carbon monoxide to water, among other findings. According to these measurements, the comet core was losing five times more water than carbon monoxide (for comet Halley in 1986, this ratio had been six to one). In addition, they found formaldehyde, carbon monosulfide, and iso-hydrogen cyanide (in this molecule, the sequence of atoms is switched; as this compound is rather unstable, it had previously been observed in cool interstellar gas clouds). Radar observations with American radio telescopes yielded a diameter of between one and three kilometers, significantly less then for the core of comet Halley.

X-Rays from a Comet

Another result made the headlines: On March 27, ROSAT discovered unexpectedly high radiation from the vicinity of the comet. The x-ray image showed a crescent-shaped area of radiation whose maximum intensity was about 30,000 kilometers in front of the comet core. This was particularly surprising, as in general, x-rays are emitted by very hot matter, while comets consist largely of frozen gases that evaporate as they approach the sun and begin to glow as a result of excitation. Correspondingly, scientists at the Max Planck Institute for Extraterrestrial Physics in Garching assumed at first that they were dealing with some kind of fluorescent glow from the coma, excited by the x-rays from the sun. However, quantitative calculations showed that the amount of gas in the coma was insufficient to explain the measured intensity. About a billion metric tons would have been necessary–between a quarter and one-half of the comet core's total mass. Collisions between dust particles of the coma and interplanetary matter (which is responsible for the zodiacal light) were ruled out as well, as they could only account for an x-ray intensity a hundred times fainter. The most probable explanation involves the conversion of kinetic energy of the solar wind to heat as it hits the coma. The energy provided by the solar wind is at least a thousand times higher than the intensity measured in the x-ray part of the spectrum.

Inspired by these unanticipated discoveries, scientists at MPE scoured the data from the ROSAT survey in search of indications of x-ray emissions from previous comets. They found three more x-ray-active comets, which had passed accidentally through ROSAT's field of view between July 1990 and January 1991: C/1990 K1 (Levy), 45P/Honda-Mrkos-Pajdusakova, and C/1990 N1 (Tsuchiya-Kiuchi), all of which when recorded had been less than 300 million kilometers from the sun. Comet Tsuchiya-Kiuchi had been seen several times by ROSAT, with a total exposure time of about eight minutes; its x-ray emission came from an area of about 470,000 kilometers.

Because of the comparatively long exposure time, the ROSAT PSPC data of the comet could also be used for a first attempt at x-ray spectroscopy. It turned out that the energy distribution can be modeled quite well by thermal bremsstrahlung, which occurs when free electrons are diverted from their original trajectory in the vicinity of positively charged ions. The spectral characteristics indicated a gas temperature of about 4.6 million K. An even better fit to the energy distribution can be achieved by adding x-ray fluorescence of oxygen nuclei with a wavelength of 2.36 nanometers. However, fluorescence of oxygen and carbon atoms alone as an explanation for x-rays from

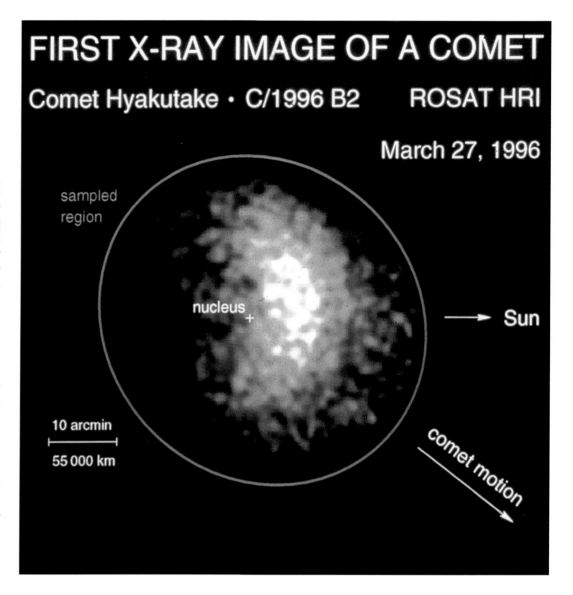

FIRST X-RAY IMAGE OF A COMET

Comet Hyakutake · C/1996 B2 ROSAT HRI

March 27, 1996

sampled region

nucleus +

→ Sun

10 arcmin

55 000 km

comet motion

The first x-ray image ever of a comet was unexpectedly obtained with ROSAT.

comets can be eliminated.

The x-ray intensity measured by ROSAT, 0.1 to 2.4 keV, for C/1990 N1 is comparable to that of comet Hyakutake, even though Hyakutake was optically about fifteen times brighter. This is further evidence that the source of the x-rays was not the comet activity itself but the interaction of evaporated comet gases with the solar wind.

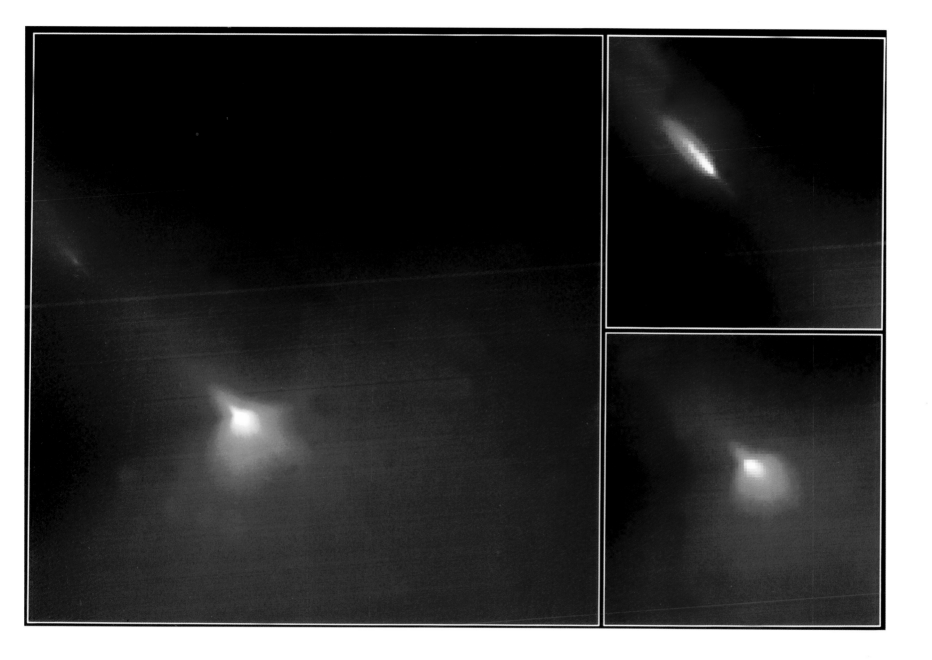

The Hubble Space Telescope took
images of the core region of
comet Hyakutake on March 25,
1996, shortly after several small
ice fragments had broken off;
these fragments could be seen for
several days as mini-comets of
their own, including a tail (top left).

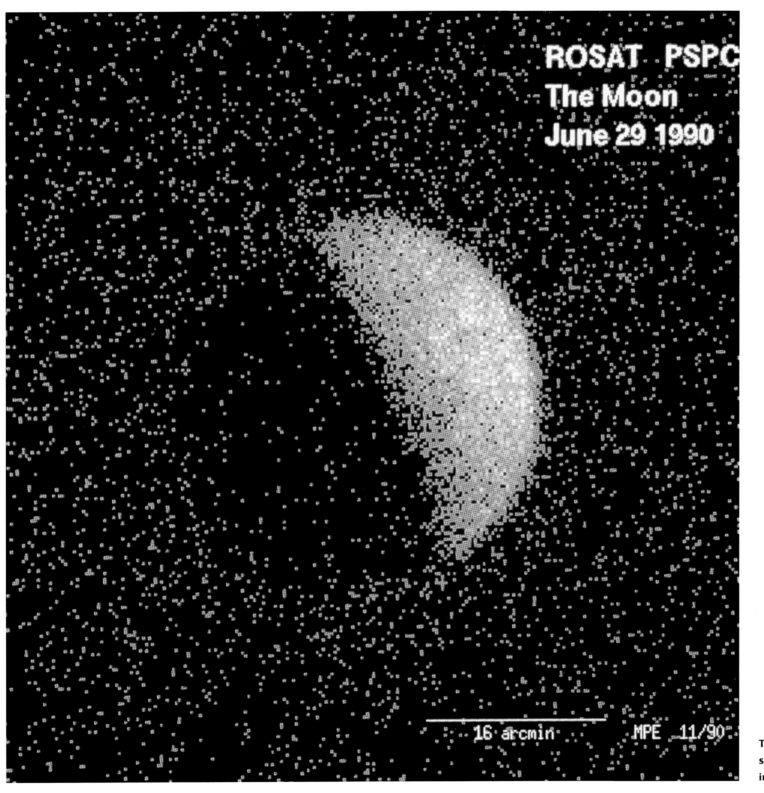

The moon, lit from the side by the sun, shows its phase in this x-ray image.

X-Rays from the Moon

Riccardo Giacconi had tried to observe x-rays from the moon with his rocket experiment in 1962, but his detector did not register anything from that direction. Instead, Giacconi discovered the first cosmic x-ray source outside our solar system, Scorpius X-1. Twenty-eight years later, ROSAT provided proof that the moon reflects not only the visible light from the sun but also solar x-rays. At the time of the observation, on June 29, 1990, the moon appeared close to its first quarter, so that its right half was illuminated by the sun. The image shows clearly that the left, dark, half of the moon shades the x-ray background radiation from the deep universe. But it still shows a few x-ray photons in this area, whose origin is still under discussion. It is conceivable, for instance, that energetic particles of the solar wind are deflected around the moon, hitting atoms in the lunar surface on the dark side, thus causing the emission of x-ray photons.

The Demystification of the Sky

Who has not been fascinated by the view of a truly dark sky in a moonless night, far from the disturbing lights of the cities? It is not surprising that our ancestors began to explore the world of the stars several thousand years ago, to shed some light on the darkness of the night.

For a long time, their main goal must have been to judge how much time remained until morning from the position of the stars, since the constellations appear in the same way every evening, changing only with the sequence of the seasons, the stars had been recognized as reliable timekeepers. But their regular and predictable ways may have led to the impression that they were removed from the rules of terrestrial transience.

With the invention of the telescope at the beginning of the seventeenth century, the objects of the sky became objects of scientific investigation. The Italian scientist Galileo Galilei observed dark spots on the sun and recognized that the nebulous band of the Milky Way resulted from the light of numerous stars, unresolved by the naked eye. Over time, all stars were identified as suns, and our own sun took its place as nothing more than an ordinary, very close star. In this century, scientists discovered the process responsible for providing all the light coming from our sun and the stars: They derive the necessary energy from the nuclear fusion of hydrogen into helium.

To start this process, the interior of a star must have temperatures of more than ten million degrees. Toward the outside, the temperature drops to a few thousand or tens of thousands of degrees at the stellar surface, with differences in temperature showing up as different colors – at least for brighter stars. Orange or reddish stars like Arcturus (the brightest star in the constellation Bootes) or Betelgeuse (the shoulder star of Orion) are significantly cooler, with surface temperatures of 3500 to 4000 degrees, than white or bluish stars like the brightest star in the sky, Sirius (9500 degrees) and Spica (30,000 degrees at its surface).

But not even these temperatures can generate x-rays of sufficient intensity to be observed over interstellar distances. Even on hot Spica, the intensity of the visible light is 200 billion times greater than the "thermal" x-ray emission. If it were not for the solar corona, with its temperature of a million degrees, x-ray detectors could not even find our own sun.

The images on page 11 make it clear that the x-ray sky, like the visible sky, is dominated by point-like objects, often resembling the constellations known in visible light. Even from the region of the Orion Nebula, known for its abundant star formation, x-rays emerge. Where does this radiation come from?

The Sun as a Prototypical Star

Even for x-ray astronomers, the sun is the key to understanding other stars because of its proximity. Early on they were able to detect x-rays from the outer parts of the solar atmosphere – the corona – and they gained an understanding of the origin of this radiation. X-ray images of the sun, taken in 1973 and 1974 with a small x-ray telescope aboard Skylab, were particularly helpful. They showed an x-ray corona that was surprisingly rich in structure and highly variable, with bright areas of activity and dark coronal holes.

Comparisons with optical images of the sun showed that the areas of activity corresponded with sunspots, which in visible wavelengths appear darker than the surrounding surface. Sunspots have been known for centuries. Their numbers increase and decrease over time: In 1843, the pharmacist and amateur astronomer Gustav Schwab found a periodicity of about eleven years from his long-term observations. At the beginning of the twentieth century, the American astronomer George Ellery Hale found that sunspots are connected by strong magnetic fields: They form the bases of large arc-shaped magnetic flux tubes, connecting pairs of sunspots and extending far above the visible solar surface. Their field strength reaches a few tenths of a tesla or more (1000 to 10,000 gauss), surpassing the strength of the terrestrial (and the "normal" solar) magnetic field by several orders of magnitude.

The differential rotation of the sun is mainly responsible for such strong local magnetic fields. The regions close to the equator rotate faster than those near the poles; the difference in the rotational period amounts to a week or more. Magnetic fields, which are trapped in the hot solar gas, become wound up and twisted like rubber bands. This process, which is still not completely understood, periodically causes these magnetic flux tubes to erupt out of the solar surface into the corona as large arcs.

Naturally, these magnetic flux tubes are not empty: Hot gas is not electrically neutral, because frequent collisions between atoms at high temperature deplete their envelopes of electrons. Physicists call this a partially or fully ionized gas or plasma; it becomes an electrical conductor and is therefore subject to electromagnetic influences. Hot solar plasma is extracted from the sun by these magnetic flux tubes, and in this process is heated even more until it begins to emit x-rays.

What Heats the Corona?

For a long time the means of additional heating of the corona were unknown. Only a few years ago, two rather different explanations existed, and solar astronomy alone could not decide between

The Japanese x-ray satellite Yohkoh ("sun ray") obtained this x-ray image of the sun on November 12, 1991; in addition to several very hot regions, it also shows several magnetic arches, in which the hot corona gas emits a sufficient amount of soft x-rays. In the dark regions, the coronal holes, gas density and temperature are too low to provide detectable radiation. (Source: J. Lemen, LPARL).

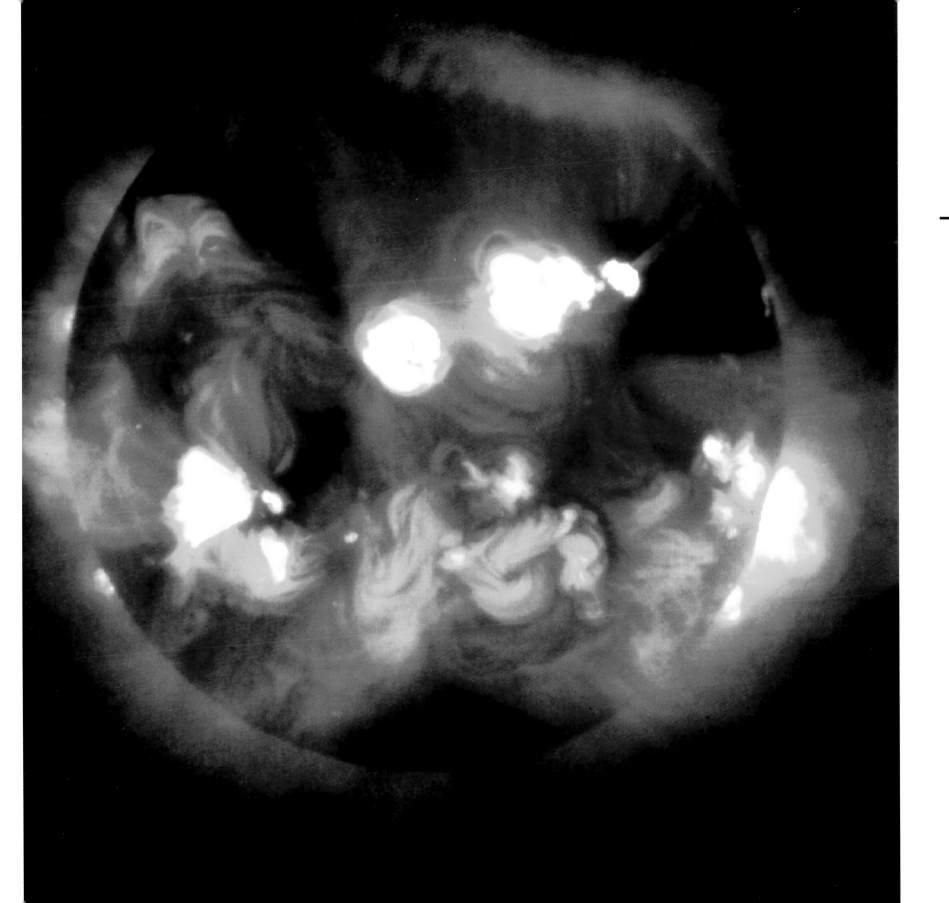

the two theories. One revolved around the possible heating of the solar corona by sound waves, a process comparable to the energy-releasing action of the surf on a coast. Breakers arise upon waves running into shallower waters, converting their energy of motion into heat.

Could not similar mechanisms be at work in the sun, where sound waves (or density waves) from the denser regions below the surface penetrate into the thinner corona? Indeed, the hot gas at the solar surface is in constant motion, like the surface of boiling water. Hot gas percolates from below, cools, and sinks back again to be reheated. This process is called convection; it appears in Earth's atmosphere as well, as warm air rises, is cooled, and sinks back to the surface.

Convection on the sun can be observed directly with suitable equipment: Detailed images of the solar surface always show a granular structure with cells of hundreds to a thousand kilometers in diameter; they exist for only a few minutes, until new cells percolate up from below. It is easy to imagine a hellish noise at the solar surface.

As an alternative, the theory of magneto-acoustic heating of the corona emerged–inspired mainly by the Skylab x-ray images of the sun. In this theory, the magnetic flux tubes act as channels, in which electrical currents are transferred from the turbulent solar surface to the corona, where they then heat the coronal gas.

Since solar physicists could not decide between the two models by observation of the sun alone, they had to wait for corresponding measurements from other stars. Here, the rich return from ROSAT, in the form of more than 120,000 x-ray sources, has brought decisive progress.

Astronomers are confined to pure observations; they cannot carry out experiments in the classical sense, such as switching off convection under the surface of a star to see if that results in a shutdown of x-rays in the corona. But luckily, nature conducts its own experiments: It makes available to astronomers stars with and without convection zones, so that they need only compare the two different types of stars.

The Hertzsprung–Russell Diagram

By the start of this century, astrophysicists had a fundamental knowledge of the physics of stars. They knew that the most important characteristics of stars – luminosity (or released energy), temperature, diameter, and mass – did not appear in all possible combinations. Rather, these parameters are linked by general physical laws. The temperature of a star is relatively easy to determine: It can be derived from its spectrum (i.e., from its color).

To classify the differences, spectral types were introduced, indicated by letters. In this scheme, the hottest stars belong to spectral type O, continuing down through B, A, F, G, and K, to the cool M stars; numerical suffixes of 0 to 9 provide even finer subdivisions. In this scheme, the sun belongs to spectral type G2 V, where the roman numeral V indicates the luminosity class of the sun (stars of luminosity class V are called "dwarfs"). If the distance to a star is known as well, its directly measurable apparent magnitude can be converted to its true, or absolute, magnitude, providing a measure of the energy released by the star.

Plotting many stars in a diagram with surface temperature (or their spectral type) as the

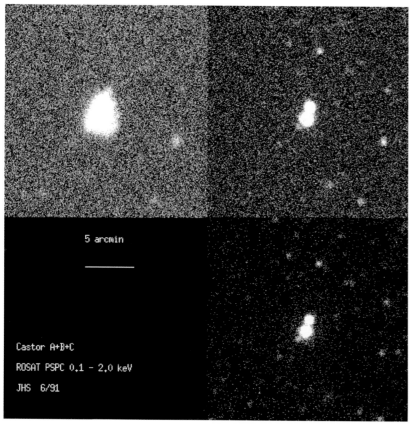

Castor A+B+C

ROSAT PSPC 0.1 - 2.0 keV

JHS 6/91

5 arcmin

Left: Castor is a system of six stars in the constellation Gemini, forming three close pairs (Castor A, B, C). The ROSAT image shows two bright, close objects: The top right object is a combination of Castor A and B, which can not be resolved by ROSAT; the bottom left object is Castor C. The ROSAT PSPC can resolve Castor C from Castor A and B particularly well at higher x-ray energies (0.9–2.0 keV, bottom right panel).

ROSAT PSPC Orion

2 degree   MPE 9,90

Right: This portion of the ROSAT sky survey shows the central area of the constellation Orion, with the three prominent stars of Orion's belt and the region around the Orion Nebula (bottom left). X-rays from the hot stars of Orion's belt (O- and B-type stars) probably originate in expanding regions of stellar wind, additionally heated by shock waves.

horizontal axis and luminosity (the energy they release) as the vertical axis, a strong concentration of data points along a thin line emerges, running through the diagram roughly diagonally from top left (hot stars with high luminosity) to bottom right (cool stars with low luminosity). This is the Hertzsprung–Russell diagram, named after the Danish astronomer Ejnar Hertzsprung and the American Henry Norris Russell, who independently discovered this relation. The astrophysical interpretation of the Hertzsprung–Russell diagram yields models for the interior structure of the stars. A star's place on the diagram is determined mainly by its total mass (and therefore its central temperature) and chemical composition (which changes with the age of the star).

When x-ray astronomers began to plot x-ray luminosity instead of optical luminosity against surface temperature, they received a major surprise. The data points were no longer concentrated along a thin line but were scattered over a much larger area. Obviously, the x-ray luminosity of a star does not depend very much on surface temperatures, and stars with equal surface temperature do not necessarily exhibit the same x-ray brightness. From these data, which were collected by ROSAT in large numbers, the scientists were able to reach an important conclusion: For sun-like stars of spectral types F or G, and for even less massive stars of types K and M, the x-ray luminosity is apparently determined by a star's magnetic activity, which depends on rotational velocity.

The situation is radically different for hot and massive O- and B-type stars, which have neither a convection zone nor a corona. From optical observations we know that these hot stars "evaporate" considerably more matter than cool stars. A mass loss of one millionth of a solar mass per year in the form of a hot, fast stellar wind (with speeds of 3000 kilometers per second) is fairly common. By contrast, the solar wind from our sun flows outward at only about 500 kilometers per second and carries only a hundred-millionth as much matter than the wind from an O-type star (just one hundred-trillionth of a solar mass per year). Scientists assume that the hot stars' strong stellar wind is heated by shock waves to such a level that it emits the observed x-rays.

Helpful Mass Screenings

But what determines the x-ray intensity of solar-type stars? To answer this question, astronomers had again to rely on statistics from a large sample of observations. Again they were aided by nature's variety.

Stars usually develop in clusters. The Pleiades, for instance, are part of such a cluster. Even with modest binoculars, many more than the seven stars visible to the naked eye can be seen; they belong to a group at a distance of over 400 light years. Another cluster, the Hyades, can be found at a distance of only about 130 light years.

We may assume that on an astronomical time scale, all members of a cluster were born at more or less the same time, so that we may more easily study the influence of mass, for instance, on the further development of a star. This analysis is aided by the fact that the stars of a cluster are also approximately the same distance from us, so that differences in brightness are real, and not caused by differences in distance.

Left: The constellation Taurus contains two well-known star clusters: the Pleiades (right) and the V-shaped group of the Hyades (below center).

Right: The young hot stars of the only seventy-million-year-old Pleiades cluster emit enough x-rays, several of them more than one hundred solar x-ray luminosities, to be observable over a distance of 450 light years.

In this way, astrophysicists have determined that the more massive stars in a cluster evolve much more rapidly than their less massive companions. Comparing clusters of different ages yields an even more revealing cross-section of the age pyramid of stars, allowing further conclusions about stellar evolution.

When astronomers used this method to explain the x-ray brightness of stars, they were able to draw unique conclusions from ROSAT observations of about a dozen star clusters; the ages of this representative sample ranged from about 600 million years for the Hyades down to approximately 20 million years for the very young cluster NGC 2232.*

*Object number 2232 in the *New General Catalogue*, a standard catalog of about 8000 star clusters, nebulae, and galaxies assembled in the last century, with its nomenclature still in use today.

Comparison of the solar-type stars in the various clusters shows that the x-ray brightness of these so-called G V stars apparently decreases with age. In very young clusters, ROSAT found an average x-ray brightness ten thousand times higher than that of the sun, and even stars in the Hyades cluster emitted ten to one hundred times as much x-ray energy as the 4.6-billion-year-old star in our backyard.

This result changes dramatically for lower mass K- and M-type stars: Even the 600-million-year-old Hyades stars of spectral types K and M do not show a significant drop compared to the members of the α Persei group, with an age of only 50 million years.

But independently we can state that young stars generally exhibit considerable x-ray brightness, and this fact is now used by astronomers to

Scan prior to flare   Scan during flare

1 degree

aid identification. Since young star clusters are located predominantly in or very close to the plane of the Milky Way, they appear against a dense background of stars, so that it becomes very difficult to pick out only the stars that truly belong to the cluster. In such a case, scientists are frequently faced with a difficult decision: If they take doubtful candidates into account, they run the risk of distorting the results with background stars, for instance in the determination of the cluster's age; but playing it safe and discarding all doubtful members may leave them with not enough stars to derive useful statistics.

Using strong x-ray brightness as proof of membership, ROSAT was able to immediately identify 110 sources as members of the 30-million-year-old star cluster NGC 2602, only 24 of which were identical to previously known cluster members. Additional spectroscopic observations showed that the large majority of them indeed belonged to that group.

The observed decline in x-ray brightness with age and mass clearly demands an explanation, and astronomers did not have very far to look. One of the classic parameters of a star changes in similar fashion: its rotational velocity. Most stars rotate much more quickly in their youth than at an advanced age, and this deceleration happens even more slowly for lower-mass K- and M-type stars than for sun-like G stars. The reason for this behavior is the gradual loss of mass (and therefore momentum) coupled with the fierce stellar winds during a star's stormy youth: The lower the mass of a star—and therefore the cooler it is—the weaker the stellar wind causing the deceleration. When the ROSAT data were compared to the rotational velocity of the observed stars, a clearly recognizable relation emerged.

Magnetic Fields Determine X-Ray Luminosity

Long before ROSAT, astronomers determined from studies of certain classes of close binary stars that magnetic activity depends strongly on the rotational speed of a star. Magnetic fields can be measured fairly easily, even over large distances.

The method for measuring stellar magnetic fields is based on an effect named for its discoverer, the Dutch physicist Pieter Zeeman. In 1896, he discovered that strong magnetic fields influenced the alignment of electrons in the shells of atoms. This occurs because electrons rotate about their axis and therefore possess a small magnetic field, which aligns itself according to the influence of an exterior magnetic field. This alignment is increasingly pronounced with stronger exterior fields, and this changes the effective energy of electrons, which in turn influences the wavelength as light is emitted—the corresponding spectral line appears to be slightly shifted. Since the aligned

Violent radiation outbursts in the x-ray range are a typical characteristic of active stars. It was only because of such an outburst that the x-ray signal of the Pleiades cluster member HII 2034 could be observed. The emitted x-ray energy surpasses that of the sun by a factor of 10,000.

electrons can rotate in two different directions, the outer magnetic field causes either a reduction or an increase in the emitted energy; integrated over many electrons, this makes the spectral lines appear to be widened, or even double or triple, depending on the direction of the line of sight relative to the magnetic field.

While it is difficult and time-consuming to obtain spectra from faint stars with the necessary resolution, astronomers have discovered many stars that appear to have strong magnetic fields. The first magnetic star was discovered by the American astronomer Harold Delos Babcock in 1948.

Scientists soon realized that magnetic stars like the sun undergo more or less periodic changes, and not only in magnetic field strength: Quite often the changes in the magnetic field are correlated with variations in brightness, leading to the hypothesis that magnetic fields are linked to (perhaps very large) star spots on these objects. This was confirmed during the eighties, when indications of star spots were found in the spectra of a number of such stars.

Many magnetic stars can be placed into one of two categories of variable stars, named after their prototypes: Alpha-Canum-Venaticorum stars and AM Herculis stars. The scheme for this nomenclature goes back to early variable-star observers. They cataloged their objects with a combination of two letters and the genitive of the (Latin) name of the surrounding constellation, unless they were among the brightest stars of that constellation, so that they already had a proper name or were indicated by a Greek letter depending on their brightness. Alpha Canum

Venaticorum is therefore the brightest star in the constellation Canes Venatici (the hunting dogs), while AM Herculis is a fairly faint star in the constellation Hercules.

Canes Venatici also contains the prototype of another class of variable stars, characterized by pronounced changes in their magnetic fields: RS Canum Venaticorum. Every 4.8 days, the brightness of this star declines by more than a magnitude for several hours. This highly regular behavior, together with the shape of the light curve and the analysis of its spectrum, makes it clear that two stars are circling each other in such a way that, from our point of view, they eclipse each other again and again. This kind of system is called an eclipsing binary. The cycle of the light curve corresponds to the orbital period. Stars orbiting each other in only a few days have to be comparatively close together and so may exchange matter. Hot gas from one component – mostly in the form of a stellar wind – streams to the other star, where it is heated and emits soft x-rays.

ROSAT observed a number of these RS Canum Venaticorum stars, among them the system AR Lacertae in the constellation The Lizard; here, the brightness drops every two days by little less than a magnitude as a result of the eclipse. From spectroscopic studies we know that in this system, the two stars orbiting each other are both larger and more massive than the sun. One is about one-and-a-half times as large, the other almost three times, yet they are only about four-and-a-half solar diameters apart. Interestingly enough, their mutual occultations are hardly noticeable in short-wavelength x-rays. This leads to the assumption that gas at a temperature of 15 million degrees,

which must be the source of the x-rays, occupies more space than either of the two stars and may envelop both of them.

In contrast, long-wavelength x-rays show the familiar light curve, dropping to a minimum whenever the larger star moves in front of the smaller one. Apparently, the x-rays from the smaller star come from giant magnetic arches, which would dwarf the comparable phenomenon on the sun. Model calculations yield field strengths ten times larger than on our central star. It is not surprising that this star shows frequent intensity bursts whenever these arches "short out." In such flares, the sudden energy bursts from these "shorts" heat the already hot coronal gas to extreme temperatures. In addition, electrons (and to a lesser degree protons) are accelerated along the magnetic field arches into lower, denser, layers of the atmosphere, where they are slowed down abruptly, producing very hard, short-wavelength x-rays.

ROSAT Registers Strongest X-Ray Burst

In August 1992, astronomers were able to observe with ROSAT an extreme radiation burst from Algol, one of the most famous eclipsing binaries. The total observing time amounted to a little less than seven days, corresponding to two revolutions of the stellar pair. During that time, the radiation increased within approximately six hours to about a hundred times its normal value and then returned to its initial level during the next 25 hours.

Analysis of the data and comparison with other observations provided interesting insights into this flare. Right at the beginning, the temperature of the gases in the outburst increased to a hundred million degrees. During the increase in intensity, the presence of heavy elements became more and more noticeable in the plasma, since the changing

Left: This image shows all x-ray sources detected as part of the ROSAT survey in the Hyades region, covering one thousand square degrees in the sky. Objects identified with known Hyades cluster members are marked with green circles.

Right: Central part of the Hyades star cluster; it is located at a distance of 130 light years and is about 600 million years old.

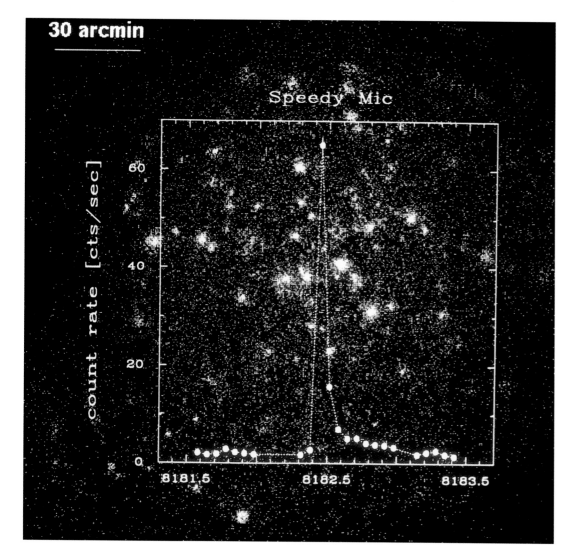

The star cluster around the bright star α Persei, with an age of 35 million years, is even younger than the Peiades and Hyades. Young stars often show strong x-ray outbursts. A particularly dramatic event is shown in the overlaid diagram: an x-ray outburst of the star HD 197890, which is, however, not part of the cluster.

conditions increasingly stimulated them to emit so-called resonance lines. Using a simplified model for such a flare event, scientists estimated the dimensions of the involved magnetic flux tube. It probably had a length of 5 million kilometers and reached to an altitude of 1.6 million kilometers (about a third of the star's diameter).

In the fall of 1992 an even stronger outburst rocked CF Tucanae (in the constellation Toucan), a binary system 175 light years distant. Within two days the x-ray brightness increased fourteenfold from its normal value, subsiding to its initial level only after nine days. In the low-energy range from 0.1 to 2.4 keV alone, in which it was observed by ROSAT, the eruption was approximately 750,000 times more intense than a strong flare on the sun. In total, it released as much energy as Earth receives from the sun in 140,000 years – enough to meet the current energy demand on our planet for more than two billion years.

X-Ray Views into Cosmic Delivery Rooms

Observations of stars in the x-ray domain also allowed insights into the very first and last phases of stellar evolution, which are invisible to optical astronomers. The birth of a star occurs in most cases behind a dense curtain deep inside extended gas and dust clouds, which absorb most of the visible light. In contrast, infrared radiation and short-wavelength x-rays penetrate these clouds almost unimpeded and can carry information about the growth of stellar embryos to the outside.

Even today, more than ten billion years after the birth of the Milky Way, new stars continue to be born. We know this because of the relation between a star's mass and its life expectancy; the more massive the star, the shorter its life. Stars like Betelgeuse, with about twenty times as much mass as our sun, can exist for only a few tens of millions of years before they leave the cosmic stage in a gigantic explosion. If such massive stars had all developed at the same time as our Milky Way, we would be unable to find a single one today. Even our sun, at an age of about 4.6 billion years, is less then half as old as our Galaxy.

Based on a large volume of observations and aided by theoretical studies, astronomers have been able to model the different phases of the birth

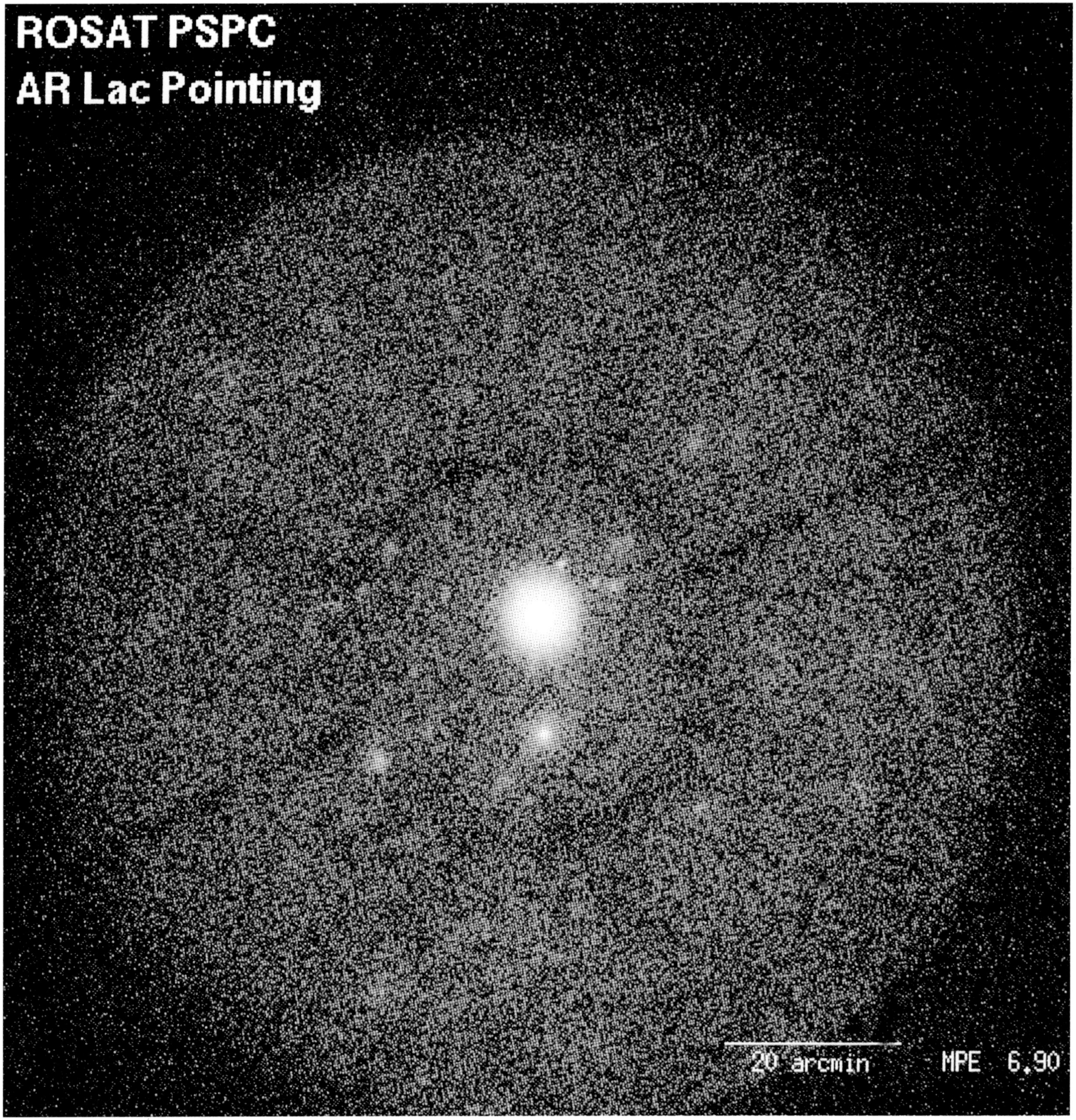
ROSAT PSPC
AR Lac Pointing
20 arcmin
MPE 6.90

The star AR Lac is an example of an eclipsing binary of the type RS Canum Venaticorum, whose characteristic feature is their enormous activity, showing up in high x-ray luminosity. During an observing sequence over more than four days, ROSAT registered the strongest radiation outburst ever observed, so that the star appears totally overexposed.

of a star. In extended, sufficiently cool gas and dust clouds between stars, small condensations develop, which over time attract more and more matter with their gravitational pull from the surrounding environment. As the density of the gas grows, temperature and pressure increase as well; the higher temperature and pressure counteract the gravitation of the cloud, and its collapse would come to a halt if atoms and molecules were not able to release part of their energy as radiation. Finally, the temperature at the center of the cloud reaches a critical value of about ten million degrees, and nuclear fusion of hydrogen to helium begins. But the newborn star is still surrounded by a dense gas and dust shell, from which matter continues to rain down onto it. While this layer prevents us from seeing the central object directly, it is heated by the star and so emits strong infrared radiation.

Since such extremely young stars almost certainly rotate very quickly, we can also expect comparatively strong magnetic activity. The stars should thus emit x-rays, which can be observed more and more easily as the thickness of the remaining layer decreases.

Indeed, astronomers were able to collect important data with ROSAT and obtain valuable insights. Before ROSAT observations, for instance, only thirty-two nascent stars were known to exist in the nearby star-forming region in the constellation Chameleon in the Southern sky. Astronomers knew twenty-five of them as classical T Tauri stars; five had been noticed by the infrared satellite IRAS; and two had been attributed to a special class of T Tauri stars characterized by very few spectral lines. T Tauri stars, a type of variable

star with irregular light curves, are considered a transient phase in early stellar development. Nuclear fusion has already started in their cores, but the contraction phase is still underway.

ROSAT identified twenty-five more members of this star-forming region, with the majority belonging to that special class of T Tauri stars with very few spectral lines. Their surrounding gas and dust shells appear to have been largely dissolved, indicating that x-ray emission from extremely young stars may limit further addition of mass, which could favor the development of low-mass stars similar to our sun.

To the surprise of astronomers, ROSAT also detected such T Tauri stars far away from the classical star-forming regions, appearing like abandoned babies where nobody had expected to find them. But it would have been almost impossible to search the entire sky for these T Tauri stars, since they are not very conspicuous in the visible part of the spectrum; their only characteristic is their variability, which is difficult to detect as part of a survey program. These isolated T Tauri stars may indicate the existence of previously unknown very small clouds of matter, in which only single stars can form, or they may have been ejected from a large star-forming region by a close encounter with another star. Additional observations over the next years should clarify this issue.

Another star-forming region is located in the constellation Cepheus. Unlike the dark clouds in Chameleon, it is a glowing gas nebula. In this region, stars appear to have formed for quite some time, and already complete massive stars provide large amounts of energetic ultraviolet radiation and x-rays; this heats the surrounding

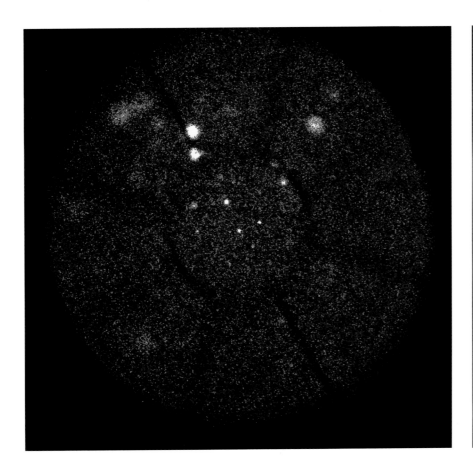

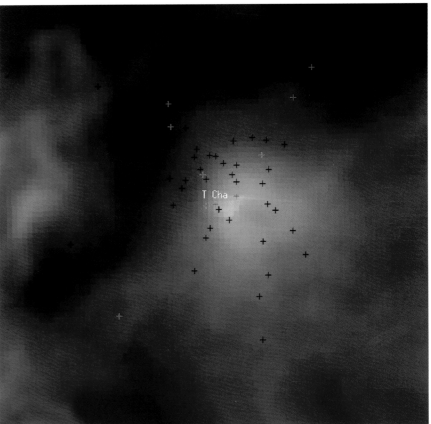

gas (predominantly hydrogen) and induces it to glow – astronomers observe a so-called H II region. (The Roman numeral II indicates that the atoms – in this case hydrogen atoms (H) – have lost their electron, that is, they are ionized.) In the *Index Catalogue* (IC), an addendum to the NGC, this region is listed as IC 1396. With a small telescope it can be seen as a reddish spot close to the star μ Cephei.

IC 1396 is almost a twin of the Rosetta Nebula and belongs to a class of old emission nebulae that are slowly dissolving. It is located in the center of a group of young, massive O- and B-type stars, the so-called Cep-OB2 association, and is induced to glow by the very young open cluster Trumpler 37. The central star of the cluster, HD 206267, is a quadruple star system, similar to the trapezium star in the center of the Orion Nebula.

While IC 1396 is about 1000 light years more distant than the Orion Nebula, the closest star-forming region, it affords astronomers a multitude of objects with which to study x-ray emission and the origin of stars. Since it appears concentrated into a smaller region of the sky because of its greater distance, it could be observed with a single ROSAT pointing.

Among other things, this emission nebula shows several small dark spots, called globules because of their spherical shape. These are local concentrations of interstellar matter, and in their interiors, new stars are probably forming. In IC 1396, several of them exhibit a bright outer layer. Ultraviolet radiation from the outside heats the surfaces of these giant gas spheres and induces them to light up. Since the interstellar gas in the "shadow" of the globules is not hit by the UV radiation, these parts of the emission nebula are fainter; only the outer layer continues to appear

Left: ROSAT image of the star-forming region in the constellation Chameleon. In the center lies T Cha, a low-mass variable star of type T Tauri.

Right: Sophisticated image analysis yielded x-ray luminosities and other parameters for a total of 54 x-ray sources in the field around T Cha. Seven of them, marked by green crosses, are also T Tauri stars.

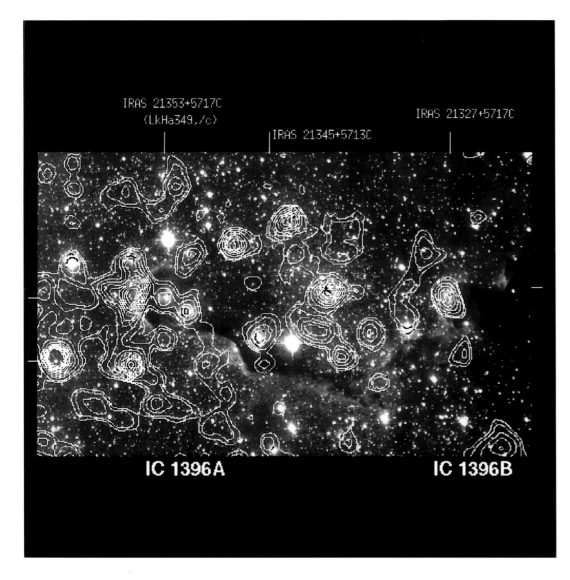

IRAS 21353+5717C
(LkHa349,/c)

IRAS 21345+5713C

IRAS 21327+5717C

IC 1396A

IC 1396B

The star-forming region IC 1396 in the constellation Cepheus contains several Bok globules – small, mostly circular dark clouds. The center of globule A contains two protostars with solar-type spectral type and age of only 100,000 years; their x-ray signature (overlaid as contours) is about a thousand times more intense than the sun.

bright, giving the shadowed parts the appearance of a comet tail.

Two apparently very young stars are located in this region; their fusion processes may have begun only about 100,000 years ago. Analysis of the optical spectrum yielded for one of them (LkHα 349) a size of three solar masses and a diameter 8.4 times that of the sun – in this case the contraction of the surrounding shell is still continuing, since "finished" stars of this mass should only be 2.5 times as large as the sun.

In June of 1993, ROSAT observed the central region of IC 1396 for more than an hour. Analysis indicated not only that several members of Trumpler 37 were x-ray sources but also that one of the two extremely young stars may be located in the interior of the globule IC 1396A (additional sources within a known object are usually indicated by letter suffixes). Other sources, however, did not show optical counterparts. Based on comparative observations in the Orion Nebula, they may be low-mass stars in later phases of development.

X-Rays from the Sirius System

In our comparison of the winter x-ray sky with its optical counterpart (pictures on pages 10 and 11) we have restricted ourselves at first to normal stars, as they can be observed with the naked eye. We have found that these stars can emit x-rays in spite of their inadequate surface temperature, as long as they have sufficiently hot stellar atmospheres or exhibit significant magnetic activity. It has also been shown that this magnetic activity is generally age- and mass-dependent, following the rule that with increasing age the magnetic activity – and therefore the x-ray intensity – decreases.

In the x-ray image, these stars appear as bluish-white dots; in comparison, the image shows only a few yellow to orange dots. The color scale has been chosen to indicate temperature differences: Hot sources, which predominantly emit short-wavelength, hard x-rays, are coded as bluish-white, while cooler ones are shown as yellow to orange. One particularly large (and therefore bright) orange object is located in the lower left corner, at the position of the well-known star Sirius. These apparently very soft x-rays do not, however, originate from the bright star visible by eye but from an object in its immediate vicinity.

The first indications of a very faint Sirius companion surfaced about 150 years ago. In 1844, the astronomer Friedrich Wilhelm Bessel had claimed, based on observations over several years, that Sirius, the brightest star in the sky, was circled by an invisible but massive companion. He had noticed that Sirius did not maintain its fixed position in the sky but rather appeared to move in a tiny ellipse, obviously caused by the gravitation of a nearby star. Almost two decades later, the American telescope builder Alvan Graham Clark found this formerly invisible companion of Sirius while testing a new telescope. Clark noticed a faint star in very close proximity to Sirius, made almost invisible by its 10,000 times brighter companion.

The true nature of Sirius' faint companion became known only when astronomers succeeded in obtaining its spectrum. It showed that Sirius B, as the companion is called, had a significantly higher surface temperature than the much brighter Sirius A.

The luminosity of a star depends not only on its surface temperature but also on its size. Since temperature differences enter into this relation to a much larger degree (to the fourth power), Sirius B should have been forty times brighter than Sirius A, were they of comparable size. Therefore, Sirius B had to be significantly smaller. In fact, it had to be 200 times smaller than Sirius A – only slightly bigger than Earth! Because of its whitish color and diminutive size, Sirius B was called a "white dwarf."

The End of a Solar-Type Star

Today we know that white dwarfs are stars similar in mass to the sun but in the final phases of their evolution. Once hydrogen as fuel for nuclear fusion runs out in their center, they first develop into a so-called red giant. Then, a strong but slow stellar wind drives their outer layers outward, while the inner parts of the red giant begin to shrink. In the end, a small white star, the white dwarf, remains – tiny, but hot.

Gas with a temperature of about 25,000 degrees mainly emits ultraviolet radiation, but the energy is sufficient for a small fraction of long-wavelength, soft x-rays. If Sirius B were not so close, its faint x-ray intensity could not have been observed even with the sensitive ROSAT detectors. In spite of this, such extremely old stars can become very interesting and revealing sources, given the right environment.

This can be the case if a white dwarf is a component of a much closer binary system than the Sirius A/Sirius B system. These two circle each other at twenty times the distance from Earth to the sun: Sirius B is about as far from Sirius A as Uranus from the sun. For one revolution around their center of gravity, this pair takes almost fifty years.

Since the period of revolution and the distance are related to each other via Kepler's third law ("the closer, the faster"), we can calculate that the components of a binary system with a period of a few hours almost have to touch each other. If two stars with a combined mass of about two solar masses circle each other every two hours, they cannot be further apart than the radius of the sun!

In such a system, at least one of the components has to be a white dwarf, as they do not require a lot of space.

Consequences of Proximity

Binary systems with such short periods are indeed known. They are usually not very bright, but their intensity shows irregular changes, with sudden outbursts that increase their brightness by a factor of a hundred. Astronomers call them cataclysmic variables; their behavior is apparently characterized by small cosmic catastrophes.

These catastrophes can be easily explained. Let us assume that the white dwarf in our hypothetical binary system has a mass of about 1.2 solar masses, leaving 0.8 solar masses for its companion; the latter would therefore be smaller than the sun and would revolve around the white dwarf at a short distance of a few tens of thousands of kilometers. Just as the Moon does on Earth, the white dwarf would cause tidal waves on the companion. However, these would be much larger, given the white dwarf's greater mass and proximity. Matter from the larger star would stream onto the more massive white dwarf.

In this context, astronomers talk about the so-called Roche limit, named for the French mathematician Edouard Roche, who investigated the interaction of gravitational fields around the middle of the nineteenth century. The Roche limit is the imaginary surface of a body where its own gravitational force and that of a neighboring object are equal. If a star expands past the Roche limit, it will lose matter to its neighbor. A mutual approach of two stars has the same consequences, since it would push the Roche limit further inward, and ultimately inside the stellar surface.

By the 1970s, several of the few then known x-ray sources had been identified as cataclysmic variables. At first, though, it was unclear where and how the x-rays were generated. Observations with ROSAT have not only significantly increased the number of known systems of this type but provided valuable details for understanding them.

Before these observations, models of the dynamics of a stream of matter had shown that the transported gas does not flow in a straight line from the star to the white dwarf, but temporarily stays in a sort of holding pattern–a so-called accretion disk, in which matter slowly spirals toward the white dwarf.

The Source of the X-Rays

In theory, there are three areas where the observed x-rays could be generated in a cataclysmic binary system. One is the hot spot where the overflowing matter hits the accretion disk; a second is where matter rains down onto the white dwarf (in both cases energy of motion is converted into heat, correspondingly increasing the temperature of the gas); finally, x-rays may be generated by outbursts in the atmosphere of the companion star.

Since the proportional counters aboard ROSAT are capable of some spectral resolution, one could try to derive information of the temperature of the source from the intensity in the various energy bands. In the case of UX UMa, a variable star in the Big Dipper, scientists determined a temperature of 1.75 million degrees from the ROSAT data. However, this information by itself did not allow

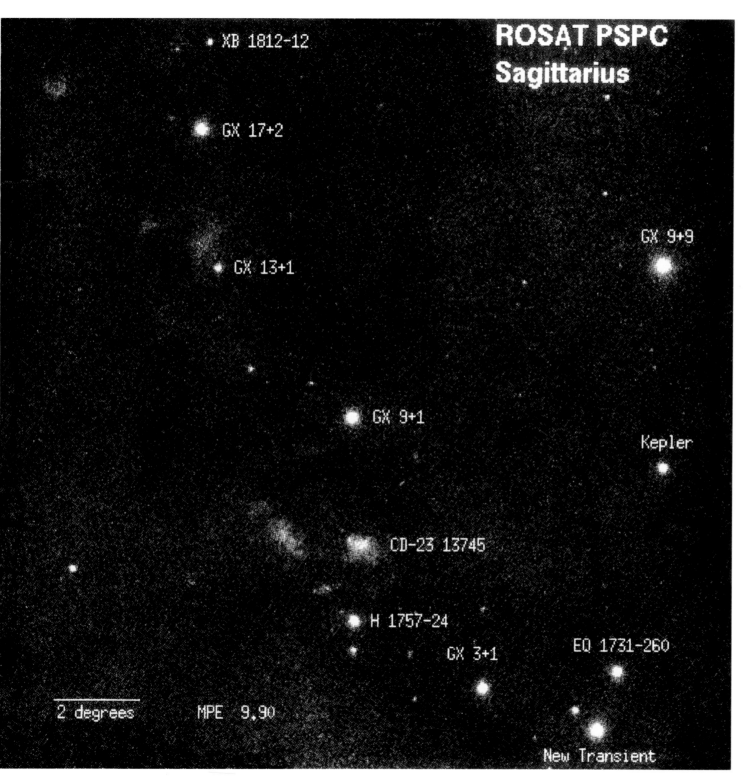

ROSAT PSPC Sagittarius

XB 1812-12

GX 17+2

GX 13+1

GX 9+9

GX 9+1

Kepler

CD-23 13745

H 1757-24

EQ 1731-260

GX 3+1

2 degrees

MPE 9,90

New Transient

The constellation Sagittarius is rich in interesting objects, such as glowing gas nebulae, star clusters, and galactic clouds, not only in the visible range, but also in the x-ray domain. Here, however, x-ray binaries predominate with neutron stars or black holes as x-ray-active partners; the diffuse radiation originates from hot gas, heated by supernova explosions.

a differentiation among the various options. The variation of the x-ray intensity over time was more informative.

X-ray astronomers did not select UX UMa by chance. The star, located halfway between the first and second stars of the "ladle" of the Big Dipper, normally shows regular light changes, appearing like clockwork every 4 hours and 43 minutes. Over a period of about 20 minutes, the star becomes fainter by about one magnitude, then returns to its original brightness. Such a light curve is typical for a binary system where we view the orbit edge-on, so that we can actually see the two stars eclipse each other.

If we see only one minimum per revolution, the two stars must be of unequal brightness or size, or probably both. Normally, the light from both stars is visible. As the larger, fainter star moves in front of the smaller but brighter star, the total intensity drops significantly; however, the occultation of the fainter star by the brighter causes little change in the total intensity of the system.

Interestingly enough, ROSAT detected x-rays from UX UMa even while optical astronomers were registering a minimum in the light curve. This could mean only that the source of the observed radiation had to be significantly larger than the occulted white dwarf itself.

This result did not fit any of the discussed possibilities, since it appeared to exclude the inner zone of the accretion disk as well as a hot spot on the outer edge of the disk. Outbursts in the atmosphere of the companion were ruled out as x-ray sources as well, since the observed x-ray intensity was too large.

But it would have been risky to draw general conclusions about an entire class of objects from observations of a single system. Observation of another dwarf nova, HT Cas, provided an additional angle. Here, ROSAT registered much harder x-rays, indicating a temperature of 25 million degrees, but also found that the x-ray signal was interrupted at the same time that the minimum occurred in the optical light curve. In this case, the x-ray source could not be much larger than the white dwarf.

In this system, the x-ray source proper had apparently been found: the inner edge of the accretion disk, where matter spiraling inward is heated the most by internal friction. The much softer x-rays from UX UMa, which also come from a much larger area of space, are "second-hand" x-rays, scattered by the particles of a stellar wind occurring in a fairly large region around the white dwarf. Energetic UV radiation from the surface of the white dwarf and the additional x-rays can accelerate gas in the vicinity into a kind of stellar wind.

Since outbursts on such stars appear at irregular intervals, they are not easy to observe. Luckily, in November 1990, during its all-sky survey, ROSAT was able to observe an outburst of VW Hyi, a variable star in the constellation Hydrus near the south celestial pole. Interestingly enough, the count rate in the energy range between 0.1 and 2.4 keV observed by ROSAT dropped significantly during the outburst and the following die-down phase, and only then returned to the level before the outburst.

X-rays in this energy range were probably only blocked during the outburst – after all, EXOSAT had observed an increase in the softer x-ray intensity during similar events earlier, which could be interpreted as multiply scattered remnants of the original high-energy radiation. These less energetic remnants would be missed by ROSAT.

From these and other observations it can be deduced that the intensity outbursts in these dwarf novae come from the region between the inner edge of the accretion disk and the white dwarf, an area smaller than the white dwarf itself. Whether, as in "proper" nova phenomena, the gas streaming onto the white dwarf causes new nuclear fusion processes to briefly ignite under the surface of the burned-out star, or whether other effects cause these outbursts, further observations may perhaps clarify.

Magnetic White Dwarfs

During the 1960s, before x-ray satellites had made their first attempts at charting the invisible sky, the hypothesis had emerged that novae and dwarf novae always occur in close binary systems with a white dwarf as one of the components. The first orbiting x-ray detectors soon found two additional subgroups of this class, containing white dwarfs characterized by significantly stronger magnetic fields. Values of 100 to almost 10,000 tesla were observed – strong enough to perturb the behavior of the entire system.

Among other things, the highly magnetic white dwarf aligns itself with the magnetic field of the companion star, so that the rotational periods of the two partners become synchronized. They show "doubly bound rotation" and appear from the outside like two rigidly connected spheres. For the first known object of this type (4U 1809+50),

known to optical astronomers as AM Herculis, showed clear spectral evidence of an extreme magnetic field. These indications occurred with a period of 3.09 hours – which corresponded exactly to the period of revolution of the two stars deduced from other observations. Apparently, the white dwarf possessed some kind of magnetic spot, which became visible and vanished again with rotation.

Such a strong magnetic fields inhibits the formation of an accretion disk. Matter streaming from the companion star is forced through wide arches along magnetic field lines directly to the magnetic spot. In this process, electrons are driven in spiral paths around the magnetic field lines. They then emit a characteristic form of radiation, whose wavelength depends on the speed of the particle and on the strength of the magnetic field. Radiation of this kind is called synchrotron radiation, since it was first observed in precursors of today's particle accelerators called synchrotrons. The typical speed of electrons under the gravitation of the white dwarf and the observed magnetic field strengths yield wavelengths in the range of visible light or the near infrared.

According to model calculations, a shock front (comparable to the frothy zone at the base of a waterfall) develops in the region where matter hits the surface of the white dwarf, that is, in the area of the magnetic poles; here, the impinging matter is heated to extreme temperatures and emits hard x-rays. Because of the dominance of the magnetic poles, these systems are called polars. Part of the radiation hits the surrounding surface of the white dwarf. This boundary zone therefore emits additional, but softer, x-rays and UV radiation.

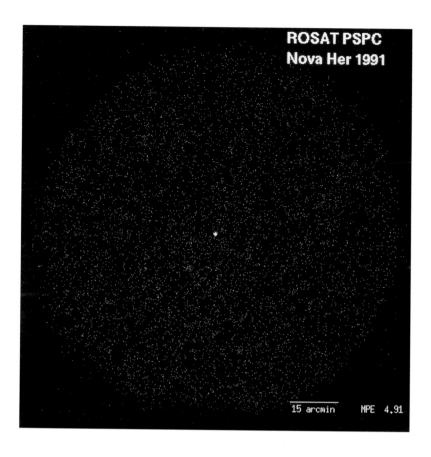

ROSAT PSPC
Nova Her 1991

15 arcmin   MPE 4.91

Novae are stars with eruptive changes in their brightness. On March 30, 1991, the nova Herculis 1991 underwent an optical outburst. Five days later, the ROSAT telescope was pointed towards the nova and found x-rays (center).

and magnetic braking effects.

The starting point is a "normal" binary system, with two components far enough apart (several astronomical units, of 150 million kilometers each) that their development is undisturbed. Given a sufficiently large difference in mass between the two stars, the more massive one will develop much faster and ultimately expand into a red giant. In this process, its extremely thin outer layer may reach the orbit of its stellar partner, so that the latter – like a satellite in Earth's upper atmosphere – is slowed down and slowly spirals in toward the core of the red giant.

During this phase, the less massive star will of course accrete matter from its partner's outer layer, but the gain in mass is very small because of the extremely low density of the gas. As this layer is heated by the second star, it simply evaporates, and this causes the braking and mutual approach to halt, leaving only a close binary system whose more massive component has meanwhile developed into a white dwarf. Later, as the less massive component reaches the red giant phase and begins to expand, it will soon encounter the now much closer Roche limit and begin to lose matter to the white dwarf – and a new cataclysmic variable is born.

Since the development of a typical AM Herculis variable relies on the magnetic coupling between the white dwarf and its partner, some maximum distance must exist below which this coupling is possible but beyond which it is not. This distance depends on the strength of the magnetic field and apparently results in a maximum period of revolution of between eight and ten hours. This could also indicate a presently unknown

This model is supported by several observations. In some systems the observed radiation periodically "switches off" during part of the revolution – at those times when the hot spot is turned away from us. In other cases the x-ray intensity reaches a minimum when the softer component is particularly bright; then we are looking straight down onto the magnetic spot, with all soft radiation from the surrounding area of the spot reaching us while the impinging stream of matter hides the hot shock front.

In the course of the ROSAT survey, about thirty new polars were discovered; this brought the total number of known systems to over fifty, with orbital periods between 80 minutes and 8 hours. This is a large enough sample to permit an attempt at reconstructing the temporal development of these bizarre pairs of stars. In general, we can assume that as the two partners slowly come closer, the system loses energy through gravitational waves relationship between the total mass of the system and the magnetic field strength.

As soon as the magnetic coupling is established, magnetic braking causes an additional loss of orbital energy, and the two stars begin to move closer to each other. This in turn results in shrinkage of the Roche limit for the companion of the white dwarf, so that the mass transfer increases.

A Mysterious Gap

As soon as the two stars move so close together that their orbit takes only about three hours, a strange phenomenon occurs. For as yet unknown reasons the stream of matter appears to stop, effectively producing a "cloaking device" for the pair. Without the hot spot, the system does not appear as bright as before, and not as blue; the characteristic synchrotron lines disappear, together with the optical intensity variations.

But the braking action and resulting convergence of the two stars continues, and consequently the Roche limit for the companion of the white dwarf keeps on shrinking, until it again reaches below the star's surface, so that the transfer of matter starts again. Now the orbital period is about two hours and still decreasing. However, the transfer of matter is no longer very productive, and so the x-ray intensity diminishes over time.

Model calculations show that this process of mutual braking does not continue indefinitely. Theoreticians rather expect a change in the inner composition of the companion star, due to the continued mass loss, leading to increasing inefficiency of the magnetic brake, so that finally, as in the Earth–Moon system, tidal friction becomes the dominant factor and begins to push the two stars slowly apart. The end result is a dissimilar pair of a white dwarf and a low-mass, dark dwarf star.

Exploding Stars

Normally, astronomers juggle time periods spanning several hundred thousand, millions, or billions of years. The development of a star from a cosmic gas or dust cloud, for instance, takes several hundred thousand years and the life of a massive star several million. Our Sun, a low-mass star, has already existed for 4.6 billion years. Therefore, changes on cosmic time scales are usually excluded from the human time horizon.

An exception to this rule is the sudden death of a massive star. It occurs within seconds and is so spectacular that its consequences can be observed over millions of light years. One of these special events surprised astronomers in February 1987, when they observed the death of a massive star in the Large Magellanic Cloud, a close companion galaxy of our Milky Way. The burnt-out core of an aged star of originally about eighteen solar masses had collapsed under its own gravity, releasing so much energy that the surrounding outer layers of the star were ejected with a speed of several thousand kilometers per second. And since the luminosity of a star depends on the temperature and size of its surface, its brightness increased rapidly with the growing diameter of the expanding layer, until it became visible even without a telescope, in spite of its distance of about 170,000 light years. A supernova had occurred – the first one visible by eye since 1604.

In 1604, at the dawn of modern science, no

telescopes or other devices existed that would have allowed Johannes Kepler or other observers to derive any detailed information about such a "new star." Today, however, astronomers command an arsenal of highly specialized equipment to extract information from this cosmic flash and its remnants.

Supernova 1987 A

The history of this unexpected death of a star turned out to be increasingly surprising the more intensively astronomers studied it. To begin with, it was a blue, not a red, supergiant (the term supergiant, by the way, does not refer to the star's diameter, but to its luminosity) that had succumbed to this sudden death. During investigations of bright stars in the Magellanic Cloud in the seventies, Nicholas Sanduleak had found a star with 80,000 times solar luminosity and a surface temperature of 14,500 degrees at this location, and he included it under the name "Sanduleak −69°202" in his catalog. Before 1987, stars with such parameters were considered quite healthy and not at all in mortal danger. It was believed that they had tens of thousands of years to live, during which they would develop into red supergiants and then end as supernovae.

It was not surprising that astrophysicists, faced with this unexpected finding, felt challenged to find a plausible explanation for the supernova death of a blue supergiant. Using computers, they began to explore the life of a massive star with models of various new boundary conditions–and found another surprise. In the Large Magellanic Cloud a star had not died prematurely, rather an overaged

object had passed away. According to the model calculations, Sanduleak −69°202 had indeed lived through the red-giant phase but had found a way to avoid death. Over a period of about 600,000 years the precursor star had lost almost half of its original mass to its surroundings–enough to reduce the strain on its interior.

While the outer layers of the star were being shed, nuclear fusion continued in its interior, generating heavier and heavier elements, as massive as iron. But at that point, no more tricks or escapes could help Sanduleak −69°202. The stellar interior had to collapse like a house of cards.

During the collapse of the already very dense stellar material, extremes of density and temperature occur. This inferno turns out to be the ideal (and in fact the only possible) breeding ground for atomic nuclei heavier than iron. A large number of radioactive nuclei are produced as well, but they soon decay.

An example of this synthesis is nickel, two steps above iron in the periodic table of elements. Nickel atoms always have two more protons than iron atoms. The number of neutrons in the nucleus, however, does not play a role in element assignment and naming. It can vary over a considerable range, changing the atomic weight. Atoms with identical numbers of protons but different numbers of neutrons therefore belong to the same element and are called isotopes of this element.

During the last days and weeks before the collapse of Sanduleak −69°202, a large quantity of sulfur and silicon nuclei developed in the interior of the star; in the five-billion-degree fireball of the supernova, they were crushed

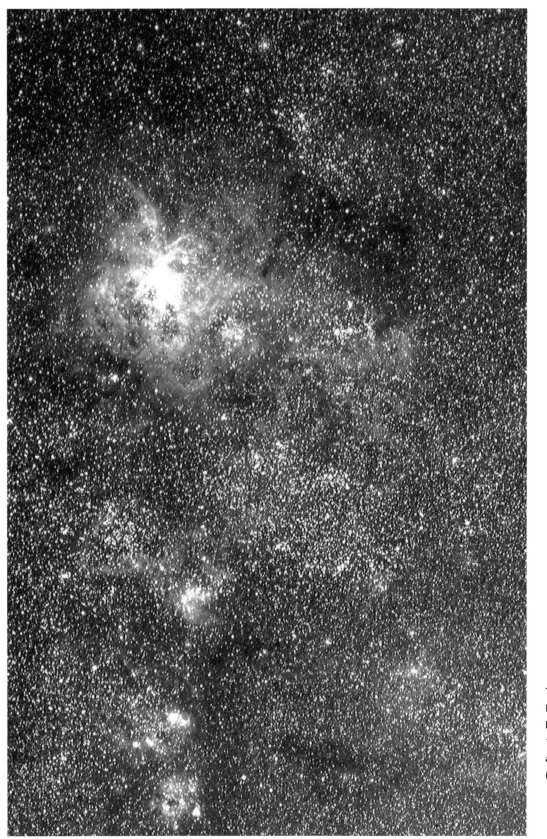

The Tarantula nebula in the Large Magellanic Cloud in visible light – before the occurrence of supernova 1987 A on the left and two days after the outburst on the right. (Source: ESO.)

against each other. This created, among other things, nuclei of the nickel isotope Ni-56 – but only temporarily, since nickel-56 is radioactive. The so-called half-life, the time it takes half of any given sample to decay, is only about six days; in that time, half of the original nickel-56 nuclei decay into cobalt-56, which is only one step above iron. This process released very energetic short-wave gamma radiation; however, it was largely absorbed in the dense shell of the star, so that it could not be detected at first outside the expanding gas shell. Only the brightness variation of the supernova in visible light indicated the fast decay of nickel-56.

But nuclei of cobalt-56 are not stable either. They decay with a half-life of about 77 days into stable iron-56 nuclei. Again, gamma radiation was emitted but still trapped by the expanding gas cloud, leading to a slight increase in the supernova's brightness. It was not until half a year after the explosion that the density of the expanding stellar remnants had decreased enough that multiply scattered – and correspondingly weakened – radiation could be observed in the x-ray band. As gamma photons traverse relatively dense matter, every collision with an electron diminishes their energy, until they are able to escape as x-ray photons. And indeed, in August 1997 the Mir–HEXE experiment registered hard x-rays from supernova 1987 A.

Toward the end of the year, gamma radiation from the decay processes could escape to the outside. Even though more than ninety percent of the originally generated nickel-56 atoms had by now vanished, scientists were able to deduce from the strength of the remaining radiation that during the explosion about eight percent of a solar mass (or more than 25,000 Earth masses) of nickel-56 was produced and had decayed.

ROSAT's First Measurements

Three years after the supernova explosion came the launch of the x-ray satellite ROSAT, capable of looking for soft x-rays from supernova 1987 A from its orbit above Earth's interfering atmosphere. This longer-wavelength and less energetic form of x-rays comes predominantly from the thermal radiation of hot gas. These soft x-rays should show either the exploding gas cloud itself or regions of nearby interstellar matter that had been overrun by the explosion and consequently heated.

The very first measurement of the satellite, the so-called "first light," was aimed at the region of the sky where Sanduleak −69°202 had suffered its recent death. From the beginning, these observations posed a difficult problem for the ground personnel at the satellite control center: Direct contact with the satellite could be established from Oberpfaffenhofen only four to five times per day – only when ROSAT was within range of the antenna at Weilheim – and then for a maximum of only ten minutes each. But supernova 1987 A is located in the Southern sky, at a southern declination of about seventy degrees. The observations had to be begun while the satellite was outside the range of the Weilheim antenna. The American colleagues were asked for help, and commands were sent to the satellite via a NASA facility near the Australian capital of Canberra.

The critical contact was established on June 16,

The first observed target of ROSAT was the supernova 1987 A in the Large Magellanic Cloud; the printout of the raw data became a souvenir because of the signatures of the participating scientists and ground personnel in the German Space Operations Center.

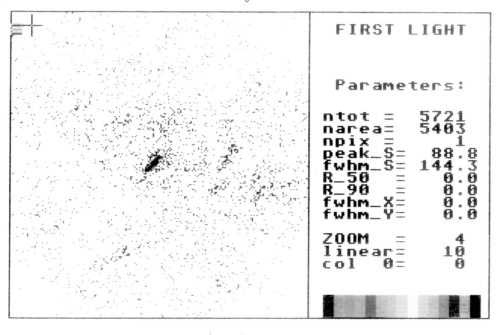

FIRST LIGHT

Parameters:

```
ntot   =     5721
narea=       5403
npix   =        1
peak_S=      88.8
fwhm_S=     144.3
R_50   =       0.0
R_90   =       0.0
fwhm_X=       0.0
fwhm_Y=       0.0

ZOOM   =        4
linear=        10
col   0=        0
```

1990, about twelve minutes before midnight local time. According to the data from the star sensors, the attitude control system had already positioned the ROSAT telescope correctly. Three minutes later the shutter of the x-ray detector opened, and the monitor in Oberpfaffenhofen began slowly to fill with little colored dots representing x-ray photons reaching the detector in orbit. Exactly at the center, a clear concentration of points developed quickly – as expected, since x-ray astronomers had earlier detected intense radiation from an x-ray binary and had cataloged the object as LMC X-1 (the brightest x-ray source in the Large Magellanic Cloud). Telescope and detector worked excellently, and the system delivered sharp images. Cheers erupted in the crowded control rooms of the GSOC and at the Max Planck Institute for Extraterrestrial Physics.

With increasing exposure time, more and fainter x-ray sources appeared: a known millisecond pulsar with the surrounding cloud from a supernova explosion; another supernova remnant, the 30-Doradus complex (a star-forming region at the edge of the Large Magellanic Cloud); and finally, a foreground star in our own Milky Way.

But at 11:59 P.M., when contact with the satellite broke off after eight minutes of exposure time, no x-ray photon from supernova 1987 A had been registered. Not until years later, during the summer of 1993, were the first soft x-ray photons observed. By then the remains of the star had moved about a third of a light year from the point of the explosion.

The Fast Supernova 1993 J

Supernova 1993 J developed much faster; it was discovered on March 27, 1993, in the spiral galaxy M81 in the Big Dipper, at a distance of about ten million light years. Just six days later, ROSAT found a strong x-ray source at the corresponding location, which had not appeared on observations half a year earlier. The reason for this difference from supernova 1987 A is related to the different history of the respective progenitor stars. While a blue supergiant had (surprisingly) exploded in the Large Magellanic Cloud, the supernova in the spiral galaxy M81 told of the abrupt end of a red supergiant.

Red supergiants deserve their name: They are so huge that if one were set in place of our sun, it would extend well beyond Earth's orbit. Correspondingly small is the escape velocity near the surface, and so a red supergiant almost loses control over its outermost layers. Thus continuous stellar wind can exist, carrying off the equivalent of several Earth masses per year at speeds of a few tens of kilometers per second. As this stellar wind is overtaken by the (supersonic) wave of the supernova explosion, it generates sufficiently high temperatures to produce x-rays. The stellar wind is sufficiently dense that this radiation is observable by the ROSAT detectors even over ten million light years – as in the case of supernova 1993 J.

The progenitor star of supernova 1987 A in the Large Magellanic Cloud had been able to convert itself into a blue supergiant after the red-giant phase; its much faster stellar wind had swept away the gas ejected earlier, leaving a growing but nearly empty bubble, so that the explosion at first

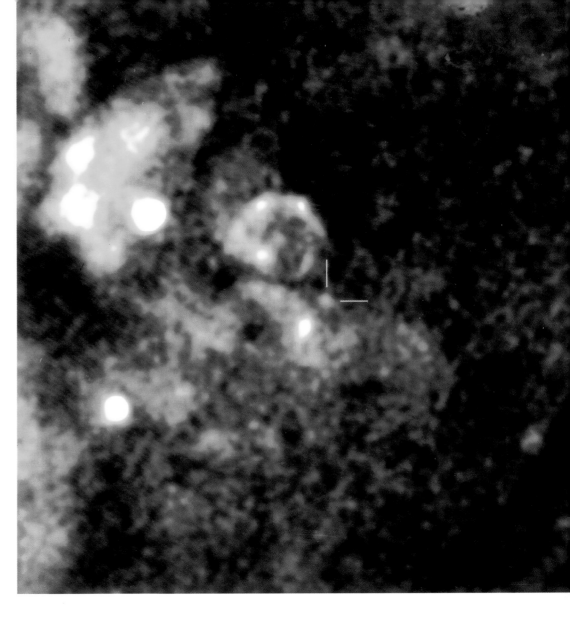

In the summer of 1993, about six years after the outburst of supernova 1987 A, ROSAT observed weak soft x-rays from this region. The cross marks the position of the supernova among other supernova remnants and the bright spots of x-ray-intense foreground stars.

Right: These two ROSAT images of the spiral galaxy M 81 were taken half a year apart and show the newly appeared supernova 1993 J (bottom right, right panel) in addition to the core region of M 81 (center) and another source (bottom).

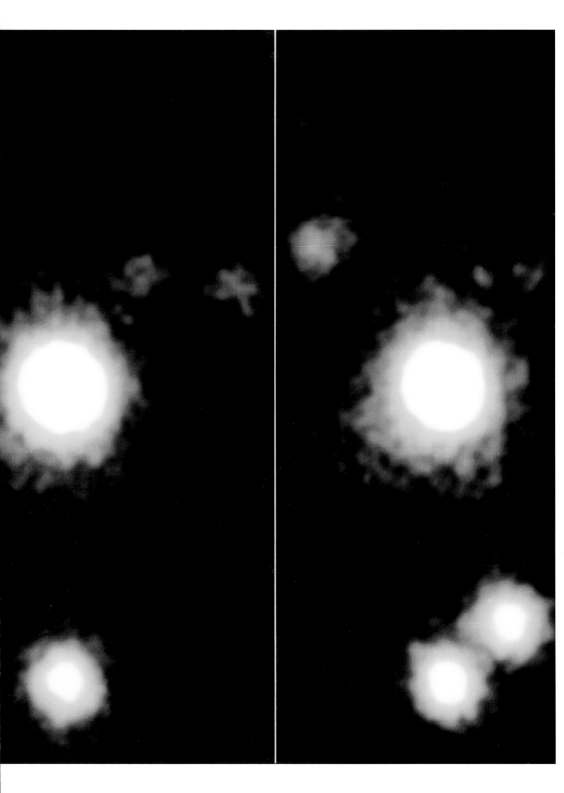

ran into near emptiness and produced no x-rays.

Galactic Supernovae

Not only in distant galaxies but even in our own, massive stars are overcome by sudden death at irregular intervals. In 1604, the last event of this type was observed by Johannes Kepler and others – this guest star, appearing out of nowhere in the constellation Ophiuchus, was as bright as Jupiter at first, fading slowly over a year until it could no longer be seen (unfortunately, the telescope was invented only several years later). Radio and x-ray telescopes show a relatively small expanding cloud at this location, at an estimated distance of about 15,000 light years.

Thirty-two years earlier, another guest star had appeared. The Danish astronomer Tycho Brahe, who would later employ Kepler as an assistant for several months, discovered it on November 11, 1572, in the constellation Cassiopeia. He observed it attentively for one-and-a-half years, attempting to measure its distance to confirm the celestial nature of this phenomenon. Since he could not determine any distance-based shifts, he had no doubts about its stellar nature. In 1573 he published his measurements and the resulting findings as a small book under the title *De stella nova* (About the New Star); as a result, these appearances of guest stars have been given the name "nova." Since then, astronomers have been waiting for the next supernova in our own galaxy. However, interstellar gas and dust clouds inhibit observations of several more distant regions of our galaxy, and a supernova occurring there may have been blocked from view.

At least one "overlooked" supernova has been confirmed after the fact and even dated, based on radio and x-ray observations: In the direction of the constellation Cassiopeia, a supernova must have occurred about a hundred years after Tycho's "new star." While the bright flash of light was shielded from view by dense dust clouds, radio astronomers observed a strong signal from this direction as early as the 1940s. In 1958, the 200-inch reflector on Mount Palomar, then the largest telescope on Earth, was able to find the optical counterpart to that source, called Cassiopeia A. It turned out to be a gas bubble, about 5 arc minutes in diameter, expanding with a speed of about 6000 kilometers per second. At an assumed distance of about 9000 light years, this corresponds to a linear diameter of approximately 13 light years, so that the beginning of the expansion (the explosion of the star) had to be about 300 years in the past. Today, Cassiopeia A is also known as a strong x-ray source.

The remains of a supernova observed by Asian astronomers on July 4, 1054 – there are no reports from medieval Europe – continue to expand at a speed of 1300 kilometers per second. The Crab Nebula, observed in all accessible wavelengths by modern astronomers at that location, seems to differ from the remnants of the Kepler and Cassiopeia supernovae. While these look like expanding, largely hollow gas shells, the Crab Nebula appears to be filled.

Spectroscopic observations show that this filling consists of very energetic electrons. When these so-called "relativistic" electrons attempt to move through the surrounding magnetic field, they are forced into spirals along the magnetic field lines and emit characteristic radiation – synchrotron radiation. Wherever synchrotron radiation appears, that is, where x-rays, radio waves, or visible light are emitted not as a result of the temperature of the source, particle accelerators have to exist that propel electrons to the required high speed (and therefore energy).

In the case of the Crab Nebula, this accelerator is actually known. It is the pulsar or neutron star that remained as the collapsed core from the explosion of the massive star. It rotates around its axis about thirty times per second, and correspondingly often the cone of radiation from its vicinity sweeps across Earth like a beam from a lighthouse.

A Stellar Explosion in the Stone Age

Only about half a dozen supernovae have been documented in historical times. The most conspicuous one, appearing in the spring of 1006 in the constellation Lupus, is said to have been a hundred times brighter than Venus at first, falling below naked-eye detectability only after about two years. These fireworks were not visible from Europe because of their far southern position in the sky (at about minus 62 degrees declination); otherwise, Europeans of that time might have started to doubt the constancy of the heavens.

Radio astronomers using their huge antennas have identified the remnants of almost 200 supernovae in the Milky Way, and about a hundred more previously unknown objects of this type were found in ROSAT observations. This significantly extended data set enabled astronomers to reconstruct stellar explosions and their history

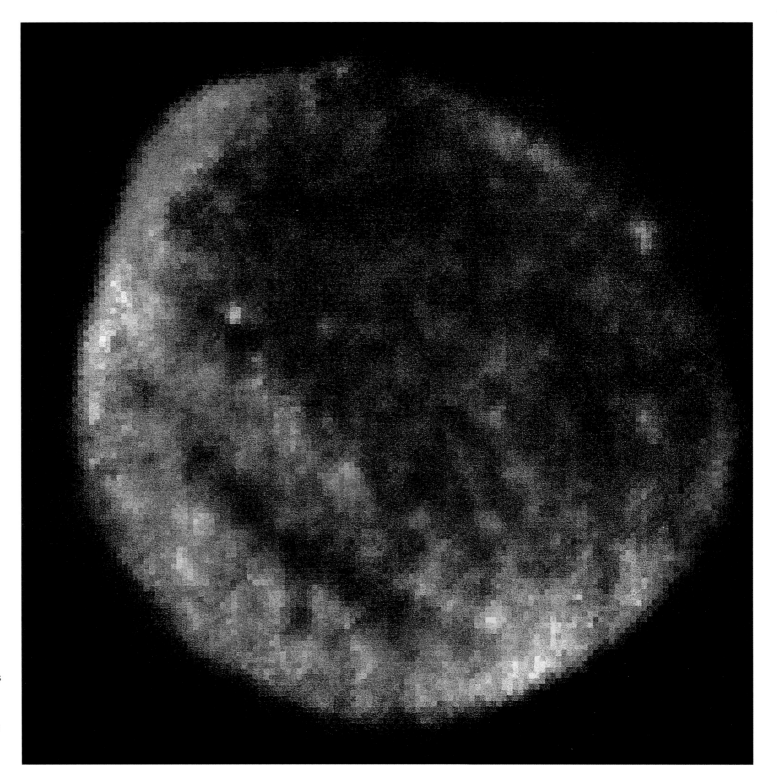

The remnant of the supernova of 1006 shows a remarkably symmetric shape, particularly regarding the two polar caps at top left and bottom right. They emit an x-ray spectrum, which is apparently generated by magnetically decelerated, extremely fast electrons (synchrotron radiation). The radiation from the remaining parts is due to hot gas, having received its high temperature from the shock of the explosion, and is now cooling by emitting radiation. (Source: University of Leicester.)

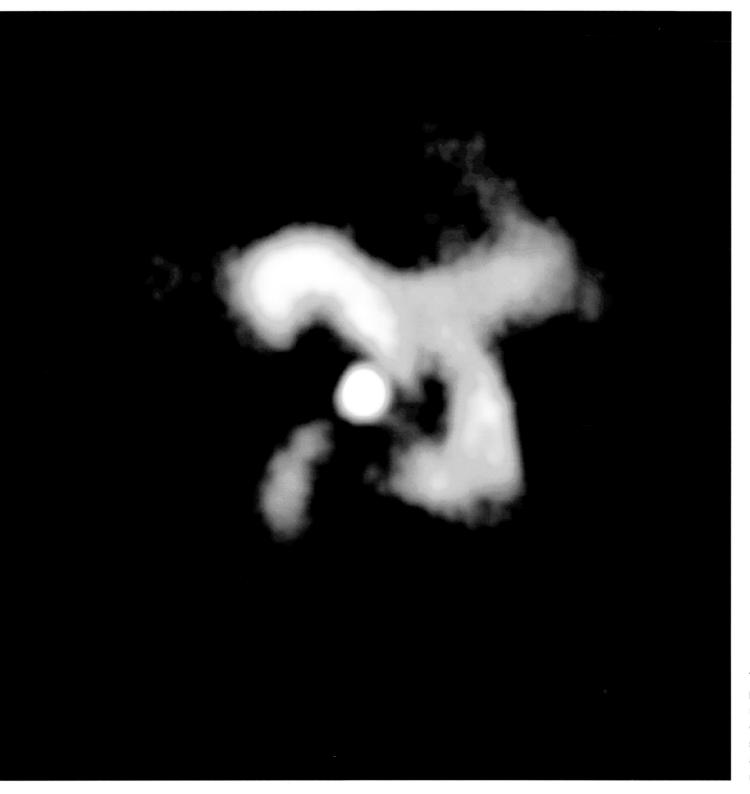

This x-ray image of the central parts of the Crab Nebula, the remnant of the supernova of July 4, 1054 (see also pages 24 and 25), shows indications of polar jets and a ring around the equator of the neutron star.

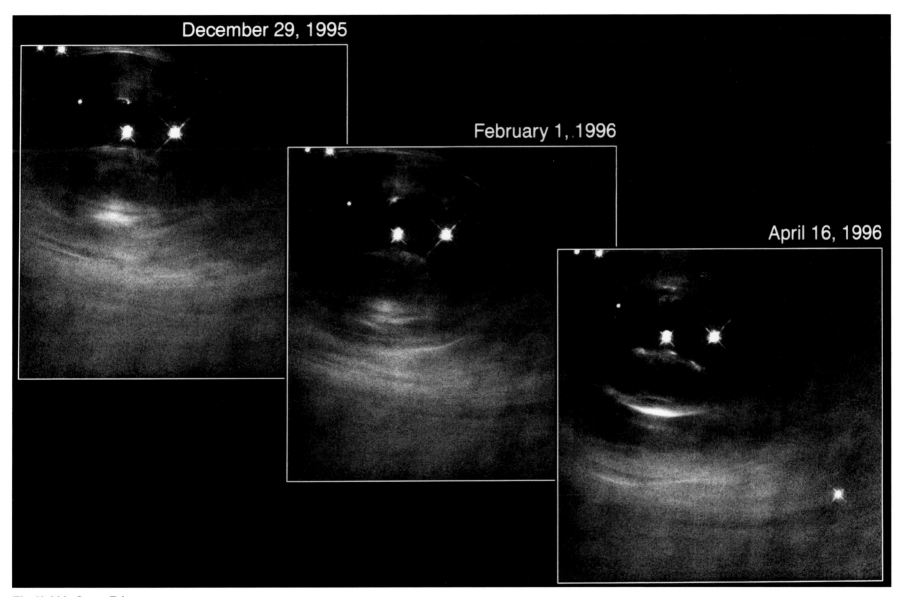

The Hubble Space Telescope was able to observe surprisingly strong changes over a few days and weeks in the appearance of the central part of the Crab Nebula; see also the HST color image on the following page.

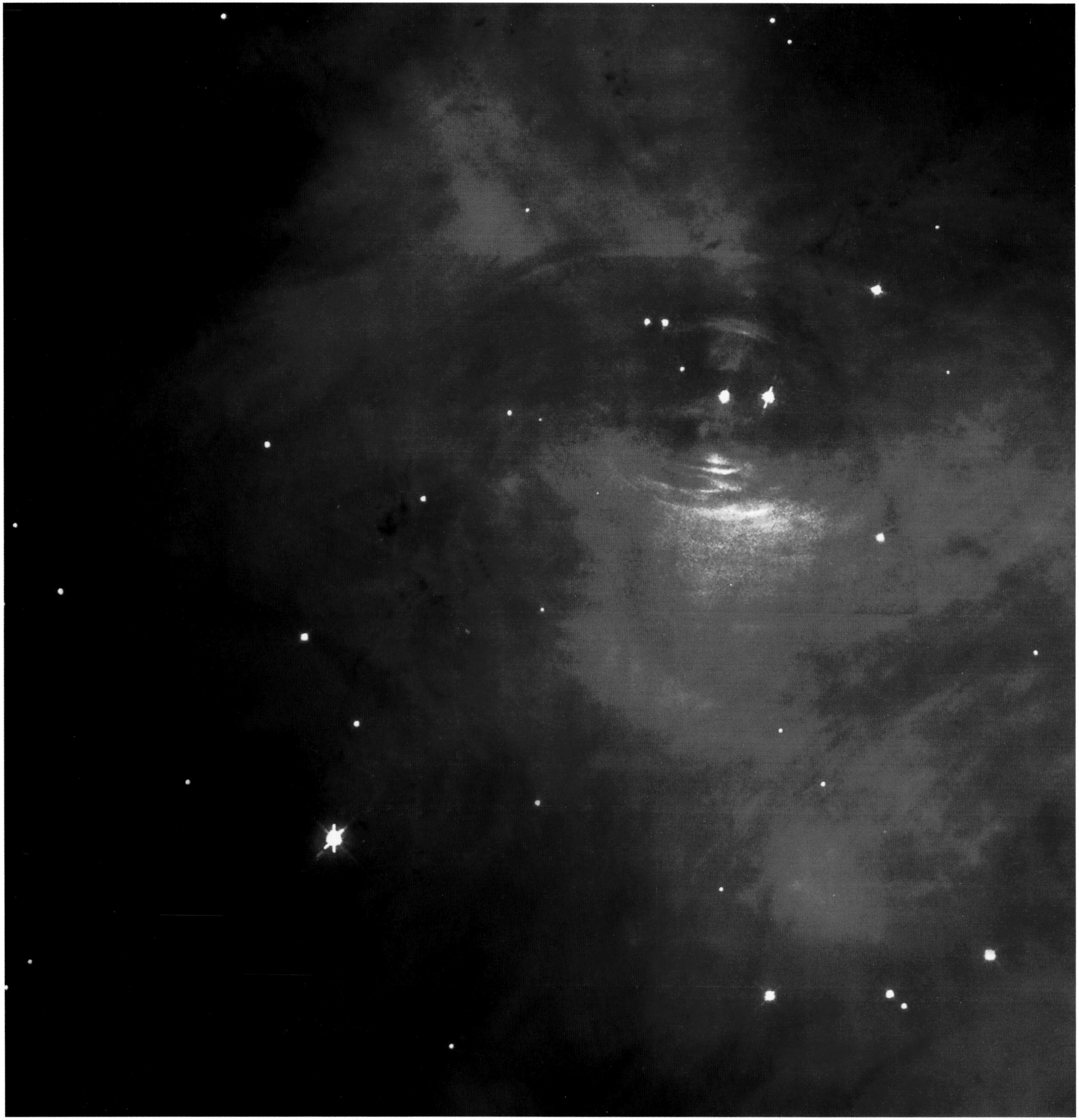

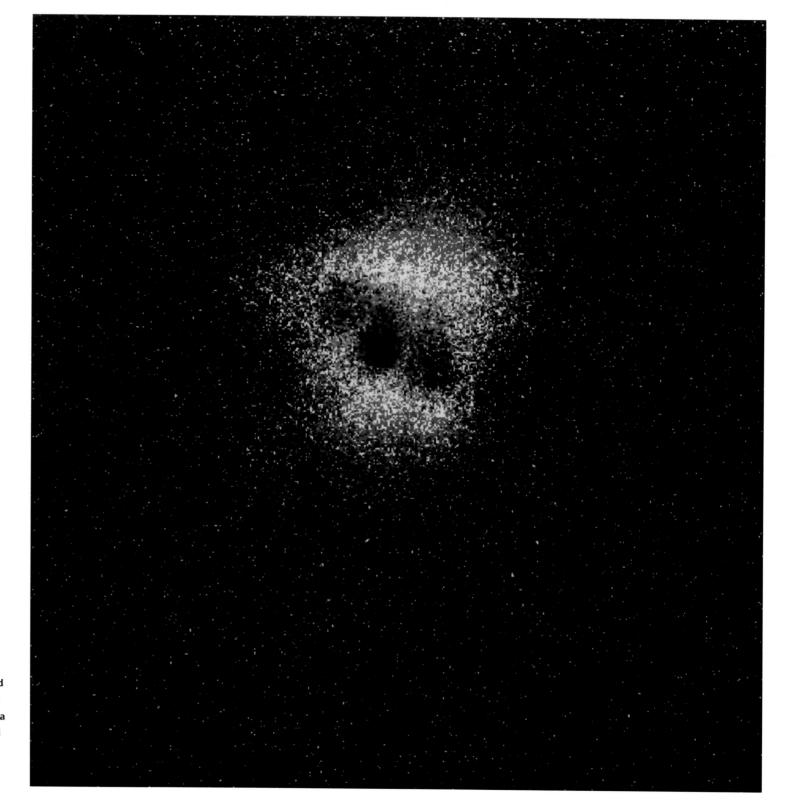

The high-energy particles emitted by the pulsar in the center of the Crab Nebula induce the supernova remnant to glow; the bell-shaped structure, which also dominates the x-ray appearance, is clearly visible.

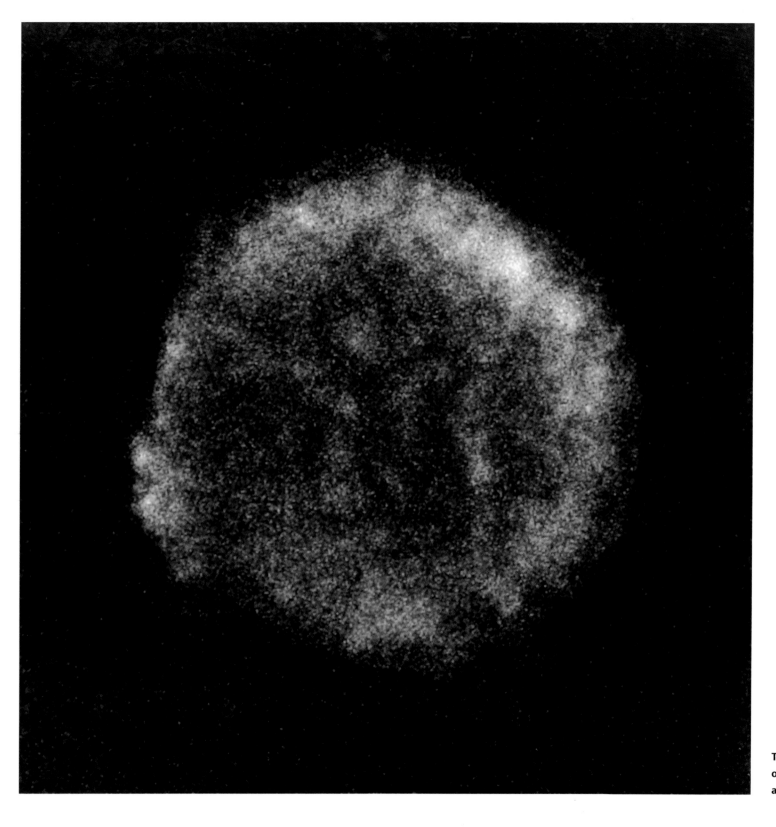

The remnant of Tycho's supernova of 1572 shows an unexpected amount of detailed structure.

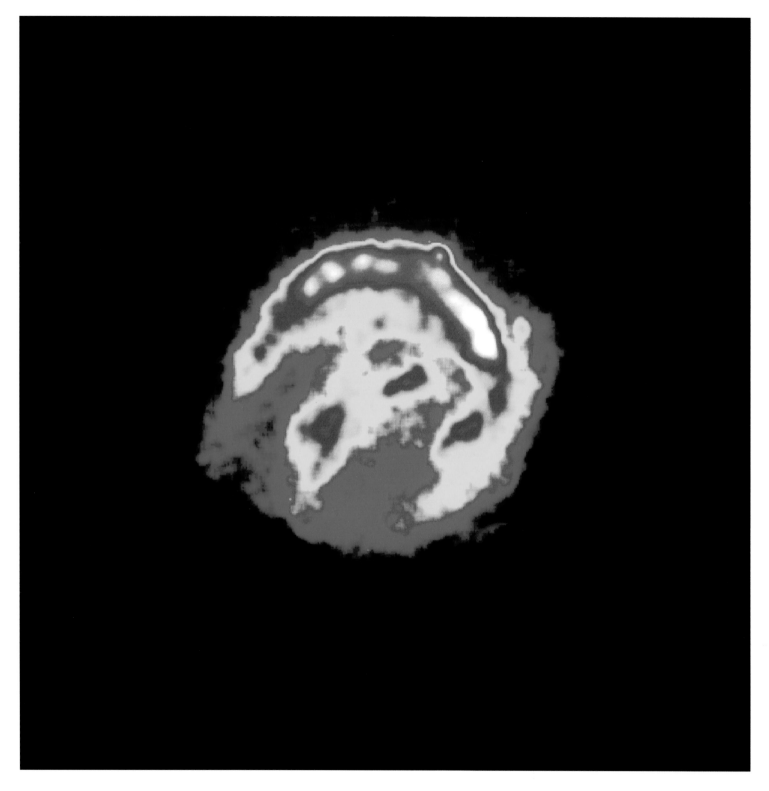

In 1604, Johannes Kepler observed a "new star," now understood to be a supernova outburst. Since then, the exploding cloud of hot, x-ray-emitting gas has reached an extent of fifteen light years. The x-ray signal is concentrated in an arc in the north and northwest, while it is much weaker in the southern parts and in the center. This ROSAT HRI image resolves structure of less than half a light year diameter.

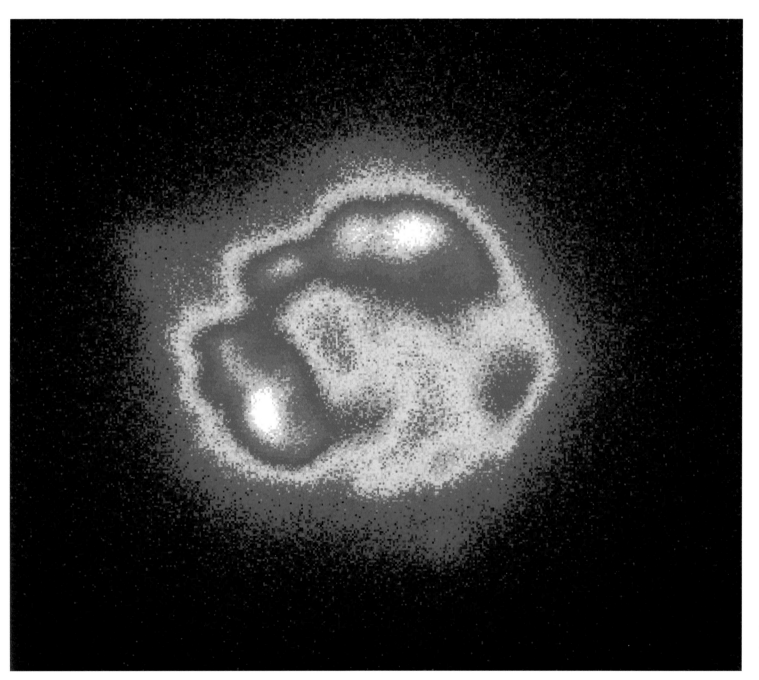

The supernova remnant Cassiopeia A is about 320 years old, but during the corresponding time period, at most a very weak supernova has been observed. The x-ray image shows strong variations in brightness, indicating large differences in matter density: White and red areas are denser than yellow, green, and blue regions.

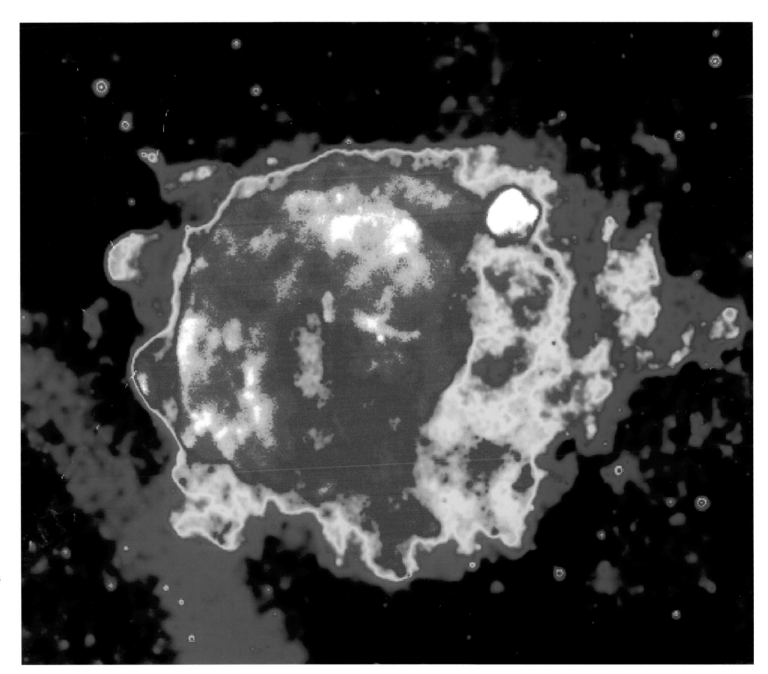

The Vela supernova remnant with a distance of only 1500 light years is one of the comparatively close clouds from such an explosion. With ROSAT scientists have found debris from the exploding star in the remnant's outer parts.

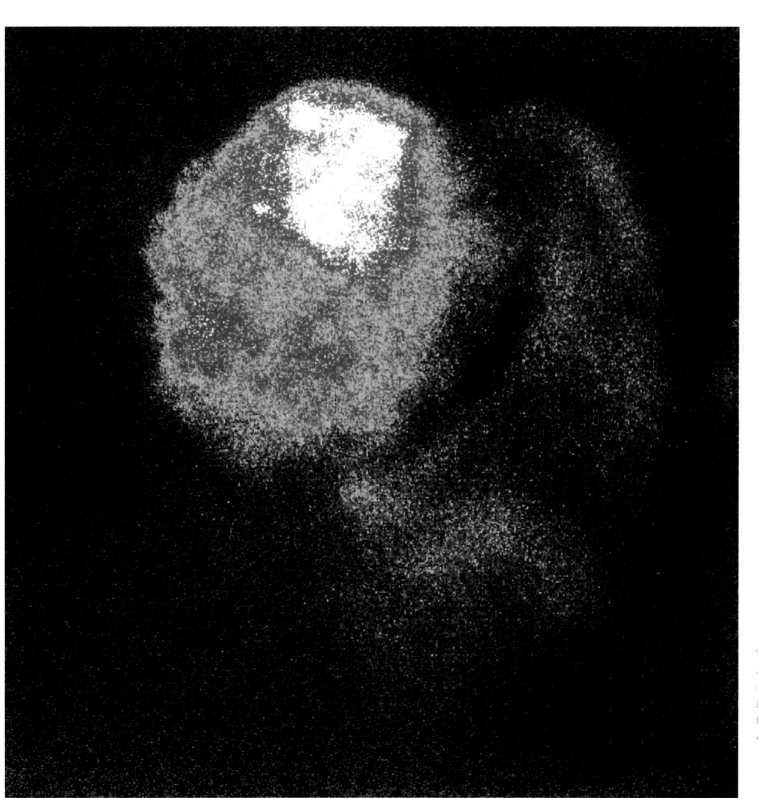

The supernova remnant IC 443 is about a thousand years old; its right part appears much darker, since the x-ray signal is absorbed by a foreground interstellar gas cloud.

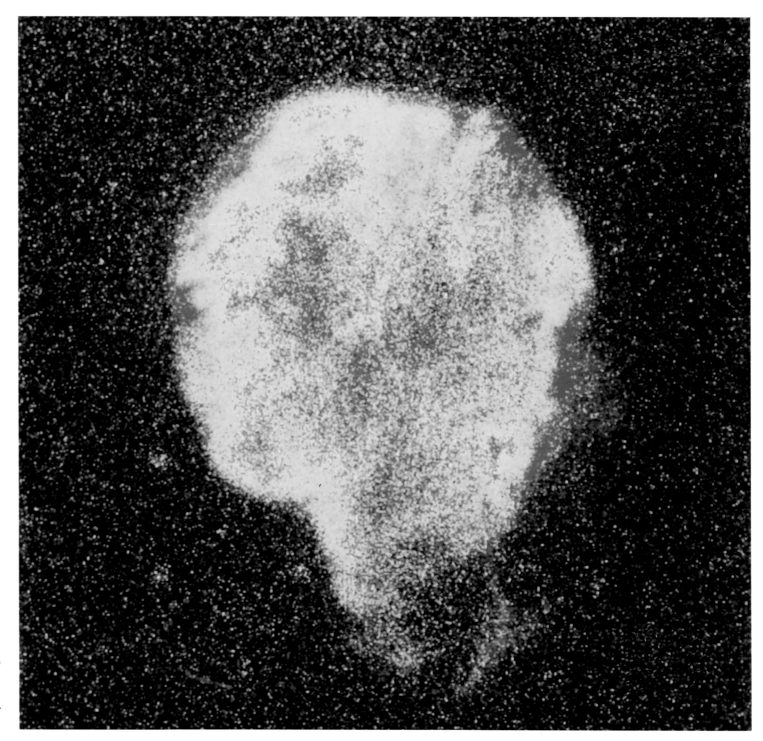

The Cirrus Nebula, or Cygnus Loop, a supernova remnant with an age of about 20,000 years, was observed with ROSAT in "color" for the first time.

much more reliably, notwithstanding the influence of local conditions, like proximity to interstellar gas and dust clouds, on the supernova remnants. Optical and radio observations by themselves cannot explain all aspects of the supernova phenomenon, as demonstrated by the ROSAT data of a supernova remnant from prehistoric times.

This sight must have been most impressive for our Stone Age ancestors, at least for those in the Mediterranean region and farther south: For several days an exploding star in the constellation Vela appeared almost as bright as a second full moon, and remained as a slowly fading speck of light over many months.

Today, only a few glowing gas fragments mark its spot in the sky, called simply Vela SNR (for supernova remnant) in the astronomical parlance. Color images of the delicate structure convey the drama of this stellar death. Interstellar gas and its distribution in the greater vicinity of the explosion served as set designer: Where the expanding remnants of the original shell encountered denser gas clouds, the increasing pressure generated sufficiently high temperatures to induce hydrogen and even oxygen atoms to glow. The blue light of the excited oxygen is noticeable as a narrow border on the edge of the expanding shell.

A neutron star as well remains from this spectacular Stone Age explosion, known to radio astronomers for almost thirty years as a pulsar with a period of about 89 milliseconds. The fast rotation of this remainder of the exploded star indicates the pulsar's relatively young age. A pulsar continuously emits energy and replenishes it from the immense potential of its rotational energy–the longer it emits radiation, the more

rotational energy gets converted and therefore the slower its rotation. Since the period of rotation can be measured extremely accurately because of its shortness, even a minute slowdown becomes evident immediately, in turn revealing the age of the pulsar; using this method, radio astronomers have derived an age of about 11,400 years.

However, the pulsar's location some distance from the center of the expanding gas cloud caused considerable headaches. If this object was indeed related to the supernova remnant, and assuming a distance of 1500 light years, it would have had to travel 36 light years from the point of the explosion, corresponding to a speed of approximately one thousand kilometers per second. This value contradicts the directly measured proper motion of the pulsar of only about one hundred kilometers per second.

The solution to this riddle came from the ROSAT sky survey. It showed that the Vela supernova remnant was considerably larger than previously assumed, and almost circular. The Vela SNR now presented itself as a glowing gas cloud with an angular diameter of 8.3 degrees, corresponding to a linear extent of 235 light years; the front of the exploding cloud therefore had to expand with an average speed of 3000 kilometers per second.

The center of this newly measured expanding cloud was now only 25 arc minutes (corresponding to a little less than 24 light years) away from the pulsar, but still "too far"; the neutron star would have had to move three times faster than actually observed to cover this distance in 11,400 years.

In addition to many known supernova remnants, ROSAT was able to detect several new ones. This ROSAT image shows the first supernova remnant detected from its x-ray signal.

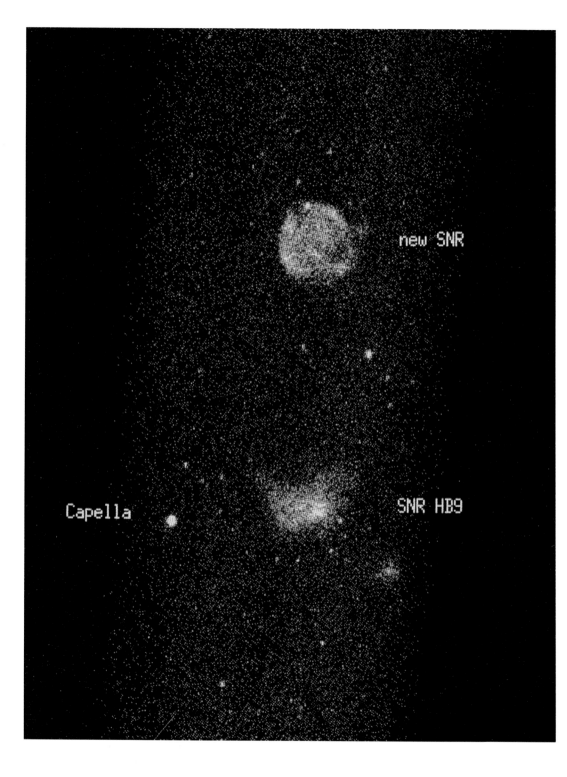

Capella

new SNR

SNR HB9

Supersonic Stellar Debris

To their surprise, scientists at MPE found conspicuous sources of radiation outside the actual
exploding cloud, partly in combination with structures reminiscent of a boat's wake. In their
unanimous opinion, these phenomena had to
have been caused by supersonic clumps of matter, ejected during the explosion of the star. This
explanation was supported by x-ray measurements from the Japanese ASCA satellite (Advanced
Satellite for Cosmology and Astrophysics), indicating that these pieces of stellar debris contained
considerable amounts of silicon, which can be produced only in the interior of a massive star. If
this interpretation is correct, then pieces of stellar debris of varying size would have to have been
preserved over several millennia. In addition, the
density of the exploding cloud could not be as
uniform as previously assumed.

This hypothesis may appear daring at first, but
it is supported by geometric considerations. If we
extend the wake-like structures backwards, they
all intersect the straight line on which the Vela
pulsar moves. As best we can determine, they
emanate from a point about 15 arc minutes away
from the present position of the pulsar (with a
plausible range of between 8 and 22 arc minutes).
This distance was traveled by the pulsar over
18,000 (plus or minus 9000) years. The resulting
minimum age includes the value of 11,400 years
derived from the slowdown of the pulsar, while
the maximum age of 27,000 years agrees fairly
well with the "geometric age" of 31,000 years, the
period the pulsar would have needed to reach
its current position from the geometric center of

the supernova remnant, given its current velocity. In the meantime, the age based on the pulsar slowdown had to be revised as well – apparently it had been determined without taking internal changes of the neutron star into account. In these changes, minute rearrangements on the surface of the neutron star lead to an increase in its rotational speed. Taking these events into account for the current rotational period of the Vela pulsar yields a revised slowdown-derived age of about 30,000 years, in very good agreement with the "geometric" age from the SNR diameter and with the "dynamic" age from the pulsar's proper motion.

The assumption that stellar fragments escape at high speed is supported by hydrodynamic models of the sequence of events in supernova explosions. Scientists at the Max Planck Institute for Astrophysics were able to demonstrate the development of large-scale turbulences in the shock wave generated by the collapse of the stellar core. These would be capable of expelling fragments containing heavy elements from the inner parts of the original star. Obviously, these fragments should be ejected in all directions (not only laterally from our point of view). Indeed, the ROSAT data show that the x-ray-luminous exploding cloud has a "pockmarked" appearance; these pockmarks may signal the presence of pieces of supersonic stellar debris inside and outside the cloud.

Pulsars in X-Rays

As spectacular as a supernova explosion may be, it does not mean that the affected star must vanish completely. In the case of the explosion of a very massive star, a so-called type-II supernova, a neutron star remains in addition to the expanding gas cloud; this neutron star is an extremely compact object of only twenty to thirty kilometers in diameter but containing as much mass as the sun, or even more. Theoretical models of the 1930s had predicted the existence of these minute stars, but they were considered unobservable, as they were thought because of their size to be incapable of emitting enough visible light.

This situation changed dramatically in 1967, when British radio astronomers discovered the then enigmatic pulsars (see page 38): Their extremely rapid and regular pulses could be explained only as coming from rotating neutron stars.

But pulsars must obey the laws of physics. To equalize the considerable loss of energy in the form of radiation, they have to draw on other resources, and the only one available is rotational energy. This transfer of energy gradually slows the rotation, which can be measured as an increase in the time between pulses; indeed, the observed values for the radiation output in the radio domain and the increase of the rotational period agree very well with theoretically derived numbers. But as the rotation of a neutron star slows over time, its accelerating force decreases as well, and this must have an effect on its x-ray intensity. It is therefore not surprising that only a small percentage of the almost 700 known radio pulsars show pulsations in x-rays – those few are thought to be no more than a few thousand years old.

A typical example is the Crab pulsar, with a rotational period of 33 milliseconds and a deceleration rate of 36.5 nanoseconds per day (1 nanosecond = 1 billionth of a second). This pulsar radiates sharp pulses not only at radio wavelengths but also in the x-ray and even gamma-ray domains, apparently due to the acceleration of extremely energetic electrons in its magnetosphere. The magnetic field strength is of the order of a hundred million tesla!

After more than five years of ROSAT observations, 22 neutron stars are known to emit x-rays; their ages, derived from other observations and from theoretical considerations, range from just under a thousand years in the case of the Crab pulsar to several billion years. This should be enough to derive conclusions about the further development of a neutron star after the supernova explosion, a subject of interest not only to astronomers but also to nuclear physicists. Many of the neutron stars discovered by ROSAT are visible in the x-ray domain simply because of their high surface temperature. Correspondingly, the relation of age and temperature can yield the mean cooling rate, and this parameter depends critically, as nuclear physicists can show, on the inner structure of the neutron star.

Despite its name, a neutron star does not contain only neutrons. The outer layer consists of iron millions of times harder than terrestrial steel. Further down, nuclei containing more and more neutrons are expected, until the neutron surplus becomes too large and they begin to "drip" from the nuclei like water from a soaked sponge. Even farther toward the center, atomic nuclei degenerate under the extreme pressure into a peculiar neutron fluid with no resistance to flow or movement. Close to the center, even neutrons cannot withstand the extreme pressure, and they begin to mutate into other elementary particles, which can be packed more densely–or even change into free quarks.

Revealing Cooling

These exotic conditions should be apparent in the cooling of a neutron star. Immediately after it forms, a neutron star possesses an extremely high temperature of several hundred million degrees and more; as it emits neutrinos, it cools rapidly and reaches the million-degree mark after about a hundred thousand years. At this time, the synchrotron radiation fed by rotational energy is much stronger than the thermal radiation from the star's surface. Observations of the Crab pulsar and the object PSR 1509-58 do not show any thermal part in the x-ray spectrum: No unpulsed, constant component is registered by the ROSAT detectors.

By the time pulsars have reached an age of several hundred thousand years, the prominent pulse structure appears significantly smoothed out. In these older pulsars, the accelerating force has been weakened enough that the pulses of synchrotron radiation become relatively weak, and the still-hot surface of the neutron star begins to dominate the x-ray spectrum. ROSAT observations of these objects show two components in the intensity distribution, a (thermal) blackbody radiation in the long-wavelength range and an additional, variable component at short wavelengths. Since this harder radiation still varies with the period of the pulsar rotation and shows the typical shape of synchrotron radiation, this component can be explained only as the last remains of radiation from the pulsar's magnetosphere.

From these objects, including the pulsars PSR 0656+14 in the constellation Gemini, PSR 1055-52 in Vela, and Geminga (an object that fooled astronomers for many years), the cooling rate of neutron stars can now be estimated. It is only necessary to derive the temperature of the "undisturbed" neutron star surface from the low-energy parts of the spectrum and to relate it to the age derived from other observations. The results are fairly clear, since the derived cooling rate agrees with the predictions of the classical theories within the achievable accuracy and manages without exotic elementary particles in the interior of neutron stars.

ROSAT also observed even older pulsars, including four so-called millisecond pulsars, rotating several hundred times per second. One of them was even discovered by ROSAT–after radio astronomers had found the first objects of this type years earlier. In these cases, old and slowed-down neutron stars appear to have received late acceleration, perhaps by mass transfer in a binary system. This is indicated by the fact that a large fraction of these millisecond pulsars are found in binary systems, where in many cases the companion has shrunk to a white dwarf. If no trace of a "donor" star can be found, it may be because it has been absorbed by the neutron star.

Another possibility may be that at the end of the "donor" phase, the stream of matter subsides; the neutron star develops into a radio pulsar and literally evaporates the former "donor." As the observed x-rays come from a spot no larger than a couple of acres but with a temperature of about two to four million degrees, it has to be assumed that the magnetic poles are still subject to continuous bombardment of electrically charged particles.

Enigma Geminga

ROSAT achieved another major success in March 1991 when it focused for a total of four hours over several consecutive days on a point in the constellation Gemini, where optical astronomers had found an extremely faint but unusually hot star several years earlier. When the data from these measurements were analyzed over the next few months, they helped to solve an almost twenty-year-old conundrum.

It had begun with an observation in the range of energetic gamma rays, which border the short-wavelength end of x-rays: All of 121 gamma photons had been received from the southwest corner of the constellation Gemini by SAS-B (the second Small Astronomy Satellite), launched in 1972. Unable to find an obvious counterpart among the dense background of stars near the plane of the Milky Way, optical astronomers had to wait until the location of the gamma ray source could be determined more accurately. Several years later, this task fell to the European gamma ray satellite COS-B, which registered more than one thousand gamma quants from this area; with them, the most likely position of the source could be determined to within 0.35 degrees. Still no optical counterpart was found–Geminga (derived from Gemini gamma ray source) remained a mystery.

Working with data from the American HEAO-2 x-ray satellite (Einstein) in 1983, Italian scientists found indications of a faint x-ray source at the location where the enigmatic gamma radiation had been observed eleven years before. Now they were able to determine the position more accurately and start a new search. But neither the photographic sky survey of the Palomar Observatory nor hour-long observations with the hundred-meter telescope of the Max Planck Institute for Radio Astronomy provided any new leads. Only with the help of the emerging CCD technology, using highly sensitive semiconductors as an electronic camera, could a small number of candidates be determined. One of them turned out to be an extremely hot object, as indicated by brightness differences in various color bands. It showed a temperature between 500,000 and 1.2 million degrees.

When ROSAT observed this source during March 1991, it found to everybody's surprise a pulsating signal. Every 0.237 seconds the x-ray intensity increased to seventeen times the average value. The enigma Geminga appeared to be a rotating neutron star from which we do not receive a radio signal, probably because the narrow radio beam misses Earth.

In the meantime, several images of the optical counterpart have been obtained, detecting fast movement relative to background stars–about three arc seconds per decade. Using the Hubble Space Telescope in 1994 and 1995, astronomers were able to determine the distance of this object. About 500 light years away, Geminga is our closest pulsar.

The unveiling of Geminga suggests that there may be large numbers of radio-quiet pulsars. The difficulty of finding them in the x-ray domain comes from the need for long sequences of observations in order to determine the unknown period. Tens of thousands of x-ray photons must be collected, and that requires extremely long observing times given the faintness of these objects. Astronomers estimate that there are a hundred million neutron stars in our galaxy alone.

Classical X-Ray Binaries

As diverse as the x-ray stars described in this chapter may be, these models do not apply to the hottest sources, which had been detected right at the beginning of the systematic exploration of the x-ray sky. Neither the energy released in the vicinity of an accreting white dwarf nor that of a rotating neutron star is sufficient to explain sources like Scorpius X-1, Cygnus X-1, or Centaurus X-3, which convert significantly more energy into radiation.

Early on it was found that these sources are related to close binary systems where matter is transported from one component to the other, as in cataclysmic binaries. The fast x-ray pulses observed from Hercules X-1, for instance, indicate that the recipient of the stream of gas must be even more compact than a white dwarf–it has to be a neutron star or even a black hole.

However, it was learned that large differences exist between the various binary systems with such compact components. Because these differences are based on the wide range of possible masses for the "donor" star, astronomers distinguish between massive and low-mass x-ray binaries: from the optical spectrum of HZ Herculis, the optical counterpart of Hercules X-1 (see page 29), we know that the temperature of the star is about 7500 degrees and that it has about two solar masses, while V799 Centauri, the optical partner of Centaurus X-3, is nineteen solar masses. While their periods of revolution are quite similar (1.7 and 2.08 days, respectively), the distances between the pairs' centers of gravity have to be quite different: about six million kilometers for Hercules X-1 and thirteen million kilometers for Centaurus X-3. But if we take the radii of the stars into account, the surfaces of the two components in each system are about the same distance apart.

For any interaction between the components of a binary system, the crucial question is whether or not they fill their Roche limit. Since the mass of HZ Herculis is only twice that of its companion while the ratio in the V799 Centauri/Centaurus X-3 system is 19:1, the Roche limit for HZ Herculis is much smaller than for V799 Centauri. Therefore, HZ Herculis can fill this limit with its diameter of two-and-a-half times that of the sun and must lose matter to its partner, while V799 Centauri remains within the Roche limit despite its twelvefold solar diameter.

But where does the matter come from that apparently rains down on the compact companion and provides the energy for the observed x-rays? Massive stars like V799 Centauri are very hot and consequently lose matter continuously in the form of a strong stellar wind, which is collected at least in part by the neutron star circling at a short distance; astronomers call this behavior Bondi–Hoyle accretion.

Difficult Search

One goal of the ROSAT mission was to discover as many new sources as possible, to allow more dependable statements based on improved statistics. Therefore, several scientists concentrated on identifying new massive x-ray binaries – not an easy task, considering the more than 100,000 sources observed in the ROSAT survey. But luckily, many astronomers develop a passion for collecting and cataloging data, and lists of extremely massive stars in our galaxy already existed.

It was only necessary to determine the intersection from both catalogs, a job that could be left to computers. However, it also had to be taken into account that the positions of the known hot stars should not deviate by more than an arc-minute from those of potential x-ray sources. Finally, the positions of about 15,000 x-ray sources from the ROSAT survey were compared with those of 15,895 massive O- and B-type stars. At the end of this lengthy selection process, and after eliminating doubtful identifications and coincidences with x-ray sources of other types, only seven objects remained.

Since their x-ray intensities are fairly faint in comparison to previously known massive x-ray binaries, these seven had not been found in earlier surveys. The ROSAT catalog may contain many more objects of this type, but their counterparts are not yet listed in optical catalogs and therefore may be much harder to identify.

Black Holes

The theory of stellar evolution predicts three

different final stages for a star, depending on its mass. A low-mass star like our sun becomes a white dwarf, stars with an initial mass from 6 to 20 or 25 solar masses develop into neutron stars; and above that range the dying star collapses into a black hole, an object that does not even let light escape its gravitational pull.

In all cases, the stars lose a considerable fraction of their initial mass in the form of stellar winds or violent explosions, so that their masses develop below the mass limits for the various final stages: White dwarfs are stable up to a limit of about 1.4 solar masses (the Chandrasekhar limit), while neutron stars above approximately three solar masses must collapse into black holes. These extreme objects of course provoke the interest of scientists, and therefore ROSAT made investigations of black holes.

At first glance it appears paradoxical to search for x-rays from an object that is dark by definition. It sounds like trying to catch a black cat in a dark room. But if such an invisible object is part of a close binary system, it should be noticeable not only from its gravitation and resulting orbital characteristics but also from more or less intense x-rays due to mass transfer. A black hole can be detected as an x-ray binary whose compact component contains significantly more than three solar masses.

At the beginning of 1996 seven such sources in our galaxy were fairly good candidates for black holes; three had been added during the previous year from ROSAT observations. The masses derived from other observations ranged from "more than 3.2" to "about 16 solar masses," so that they were unlikely to be neutron stars. With one exception (Cygnus X-1) all candidates have the appearance of low-mass x-ray binaries, where slowly expanding stars lose part of their matter across the Roche limit onto their compact partner.

In addition to mass determination, the x-ray spectrum can provide clues to the nature of the compact companion. In can be assumed that the radiation predominantly emanates from the hot, inner parts of the accretion disk. Above this disk, astronomers suspect a thin atmosphere, which is heated by turbulence and perhaps by magnetic effects. As soft x-rays penetrate this atmosphere, they are repeatedly scattered on hot electrons, taking part of their energy with them and bringing them to higher energy levels. The spectral distribution of this type of radiation drops off abruptly towards shorter wavelengths, at the point where the energy of the x-ray photons reaches that of the hot electrons, so that no further gain in energy is possible.

For black holes, this limit is suspected to be between 50 and 100 keV, corresponding to a temperature of more than 500 million degrees. These values cannot be realized in the vicinity of a neutron star. But since ROSAT does not cover this energy range, scientists have to rely on observations from other satellites.

The Particle Slingshot SS 433

In 1976, the British x-ray satellite Ariel V registered x-rays from the constellation Aquila. Radio astronomers had observed an extended supernova remnant (W50) in this area several years before, so the new x-ray source (A1909+04) was immediately identified with the supernova.

But neither radio nor x-ray astronomers knew that two optical astronomers had noticed a peculiar star in that direction: Bruce Stephenson and Nicholas Sanduleak had found a star of fourteenth magnitude (and included it in their catalog as number 433) that showed unusually bright lines of hydrogen in its spectrum. Normally, spectral lines appear as a pattern of dark absorption lines, since the cooler gases of the stellar atmosphere partly absorb the light from the stellar surface; bright emission lines always indicate significant quantities of hot gases.

When detailed spectra of this source, called SS 433, were finally taken during the summer of 1978, astronomers were in for a surprise. Aside from the known emission lines of hydrogen, other bright lines appeared, which wandered back and forth across large parts of the spectrum with a surprisingly regular period of 164 days.

Moving spectral lines appear in almost any close binary system. As long as we are not looking straight down on (or perpendicular to) the plane of the orbit of such a system, one of the two is always moving away from us while the other approaches. This causes the spectral lines of the former to be shifted toward the red and those of the latter toward the blue. This is the well-known Doppler effect, named after the Austrian physicist Christian Johann Doppler, who developed a conclusive explanation about changes in wavelength as a result of motion in the 1840s.

At the turn-around points, the two systems of lines coincide and then change the direction of their Doppler shifts. From the amplitude of the line shifts astronomers determine the orbital speed of the two stars and also deduce their masses; normally, these speeds are below a thousand kilometers per second, with the higher values occurring only with very short orbital periods of several hours.

With SS 433, the measured speeds reached over 40,000 kilometers per second, with an orbital period of 164 days. But these values could not be caused by orbital motion, because either the speed or the period was much too large, or their masses were too low. To force speeds of 42,000 kilometers per second with an orbital period of 164 days, SS 433 would have to total two billion solar masses.

It was some time before the theoreticians overcame their shock and were able to provide a model that explained this bizarre situation. But then everything fell into place: The SS 433 system contains a massive B-type star circled by a compact companion at close range, with the former losing matter to the latter across the Roche limit. As in other x-ray binaries, the streaming matter is first collected in an accretion disk. Apparently, the rotation axes of the two stars are fairly strongly tilted with respect to each other, so that the accretion disk is located at an angle relative to the equator of the B-type star; consequently, it is forced into a spinning precessional movement with a period of 164 days by the gravitational force of the B-type star. Part of the matter in the innermost region of the accretion disk is ejected in the two directions perpendicular to the disk – possibly because the high density and temperature values resulting from the considerable flow rate of one millionth of a solar mass per year cause the disk to overflow. The ejected gas streams outward with a speed of 78,000 kilome-

The x-ray binary SS 433 shows remarkable jets on both sides, in which matter is ejected at over one-fourth the speed of light.

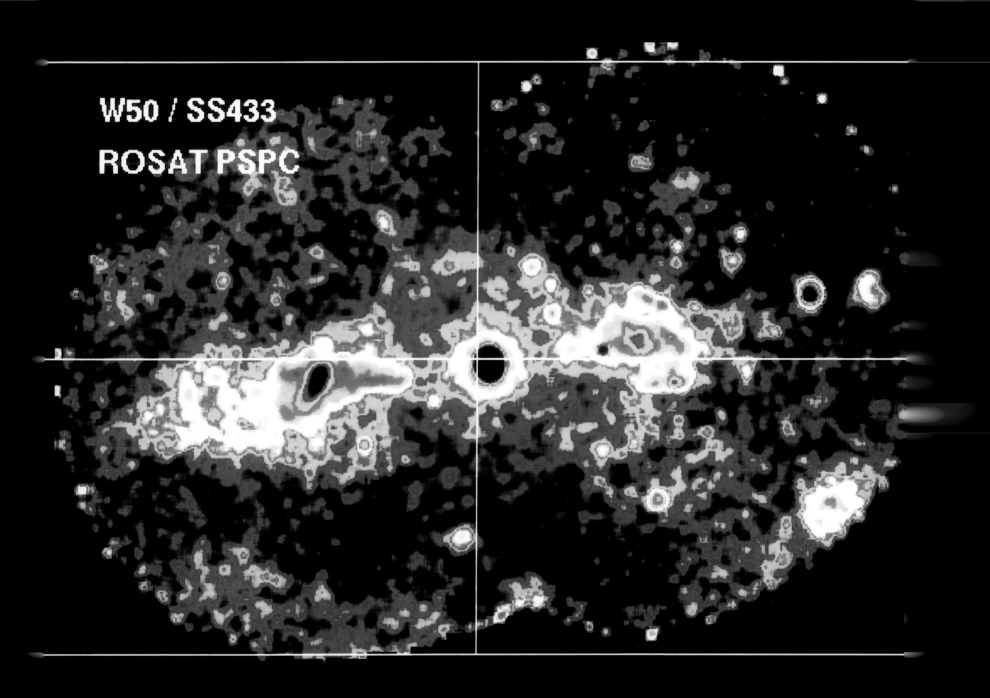

ters per second – more than one-fourth the speed of light!

Observations of SS 433 with ROSAT have shown that the ejected gas near the common base of the two jets has a temperature of at least ten million degrees. However, the actual origins of these jets cannot be observed in the x-ray domain, since they – and the compact object as well – are hidden by the precessing accretion disk. But the discovery of SS 433 has shown scientists many details about a phenomenon seen outside the Milky Way on a much grander scale – active galactic nuclei.

Another peculiarity, first observed in distant quasars, has now been found and studied in our own galaxy: speeds seemingly larger than the speed of light. In these cases we look almost perpendicularly onto a system similar to SS 433. When in such a jet a distinguishable spot moves almost exactly toward us at close to the speed of light, the small transverse component of its movement appears disproportionately enlarged by relativistic effects, giving the impression of a motion faster than the speed of light.

The first two sources in our own galaxy that showed this phenomenon had initially been noted as "hard x-ray transients." For instance, the object GRS 1915+105 (GRS indicates a source discovered by the Russian x-ray satellite Granat, while the numbers indicate the position of the source in the sky, in this case pointing toward the constellation Aquila) developed several times over the course of three years into the brightest hard x-ray source in our galaxy. Such transient sources behave much like novae in the hard x-ray range above ten keV – a fast rise in intensity followed by a slow decrease.

In the optical range these nova occurrences have been identified as the result of fusion reactions close to the surface of white dwarfs in close binary systems. This reaction sets in whenever the matter that has streamed from the companion to the white dwarf reaches sufficient density and temperature to ignite a brief episode of fusion of hydrogen to helium.

Compared to these systems, the intensity of the hard x-ray transient in Aquila increased relatively slowly. The spectral characteristics observed with the help of the Compton Gamma Ray Observatory, a satellite to explore cosmic gamma rays up to energies of 220 keV (corresponding to a radiation temperature of more than 2.5 billion degrees), indicated a black hole as the compact partner in the binary system.

From the x-ray luminosity of the matter ejected during these outbursts, the typical amount of ejected material can be calculated. It amounts to about one four-thousandth of Earth's mass, so that – with one outburst per month on average – about one hundred-millionth of a solar mass per year is catapulted from the system. The energy of motion released from the system is of the same order of magnitude as the emitted x-ray energy, which also indicates a black hole as the "engine" of GRS 1915+105. Rashid Sunyaev and a colleague had predicted this ratio in 1973 in a groundbreaking theoretical work on accretion disks around black holes.

Based on these observations, astronomers have compiled a list of sources from existing catalogs whose characteristics might be explained by such a scenario. However, no concrete indications for relativistic jets have yet been found for those;

the object Cygnus X-1 also belongs to that group. They will be examined more closely in the future.

Impenetrable Clouds

X-rays (at least the softer parts) are absorbed not only by dense accretion disks, but also by the considerably thinner but larger interstellar gas and dust clouds. Optical astronomers have observed this extinction of light for a long time, as it shows up, for instance, as many dark areas with a low density of stars in the plane of the Milky Way. Toward the end of the eighteenth century, Frederick William Herschel considered these dark areas "holes in the Milky Way" that would allow us to look into the depths of space. About one hundred years later, however, the American astronomer Edward Emerson Barnard showed that these presumed holes were only an illusion–in reality, they were huge dust clouds absorbing the light from more distant objects.

Spectroscopic observations (and, later, color images) of the dark clouds and their vicinity have shown that the color of visible light is changed as it is partly absorbed. Background stars behind dense clouds appear reddened, as the blue, short-wavelength part of the light is scattered more strongly. A similar effect can be seen during sunrise or sunset. The molecules of our atmosphere predominantly scatter the blue part of the sunlight from its straight path, giving rise to the blue color of the sky; in the morning or evening, when the sunlight hits the atmosphere at a large angle and has to take a much longer path through it, its blue component is scattered so much that only the remaining red light reaches us. The

sun appears redder and much less bright.

With the help of diffraction theory we find that interstellar extinction is caused mainly by particles with a typical size of a ten-thousandth of a millimeter (0.1 μm). These dust particles can be pure carbon (such as graphite or "soot"), or they may contain cores of silicates with deposits of organic or nitrogen compounds.

Interstellar extinction decreases rapidly toward longer wavelengths, so that in the near infrared our view through our galaxy is almost uninhibited. Toward shorter wavelengths, radiation is increasingly absorbed by gas atoms and molecules.

The absorption reaches a maximum in the area of the ground state of neutral hydrogen, around 912 angstrom. Radiation of this wavelength (and shorter) is able to extract the electron from a hydrogen atom and thus ionize hydrogen; but since hydrogen atoms make up the largest part of interstellar matter, radiation of this wavelength does not make it very far in space. As a result, only very bright and close objects can be observed in our galaxy with telescopes for the extreme ultraviolet (EUV), at wavelengths below 912 angstrom (or energies above 13.6 eV). In the x-ray domain the absorption decreases again and is barely noticeable above 10 keV (or at wavelengths below 1.24 angstrom).

In addition to this absorption, scattering on dust particles also plays a role in the x-ray range. Thanks to the high sensitivity of the ROSAT instruments, sequences of observations could compare the optical dark clouds of our galaxy with those appearing in x-rays. They show striking agreement, even if the contrast of the x-ray images

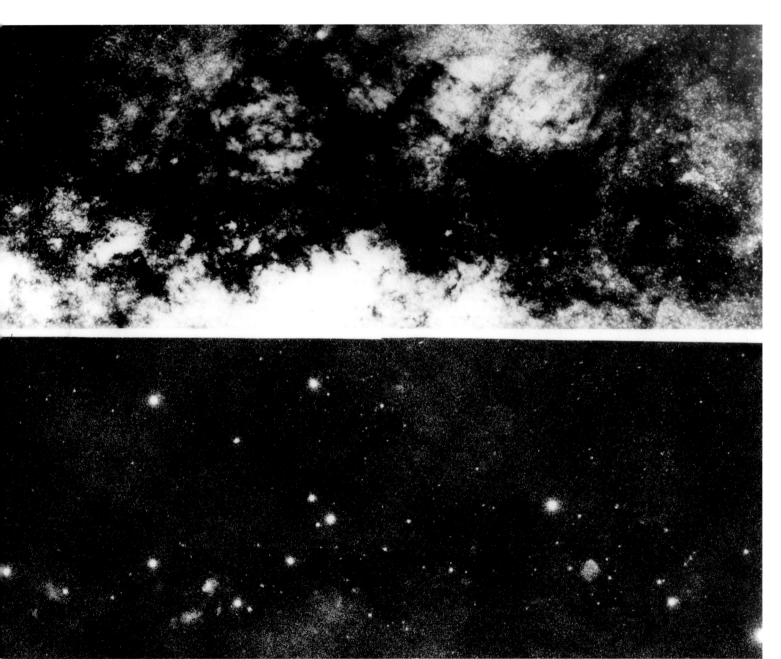

The band of the Milky Way, corresponding to the plane of our galaxy, is interrupted in many places by dark clouds of interstellar dust and gas (top). A comparison of the same region in an x-ray image shows that the propagation of soft x-rays is also influenced by these clouds.

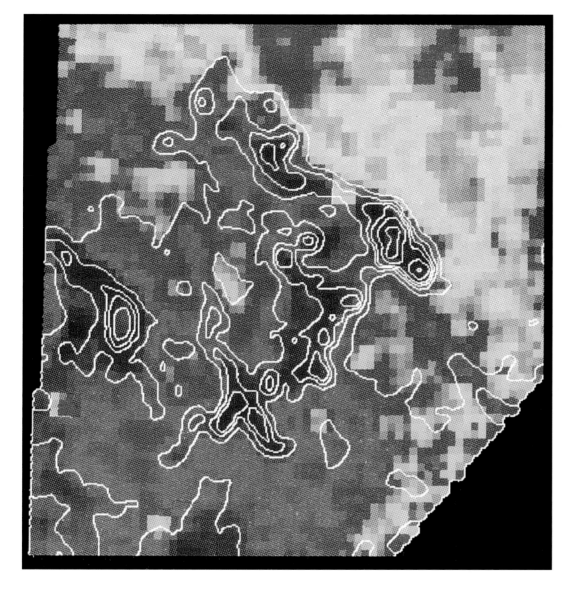

The superposition of x-ray intensity (in color) and infrared measurements (contours) obtained with the IRAS satellite shows the shadowing of the diffuse x-ray component by dense interstellar dust clouds, which are visible in the infrared; this indicates that at least part of the x-ray signal has to come from the galactic halo.

is not as high.

As the x-ray absorption over the energy range observable with ROSAT (0.1 to 2.4 keV) decreases by a factor of one thousand, the ROSAT instruments are in an excellent position to determine the amount of optically effective matter (the so-called column density of interstellar matter).

Measurements of scattering halos from dust had been mentioned since 1965; they should make it possible to derive the size, and size distribution, of the dust particles. These observations require extremely smooth mirrors, low in scattering, to avoid the masking of the small real results by large instrumental effects. The excellent quality of the ROSAT mirrors enabled scientists to make significant progress in this area as well.

It has been particularly helpful to observe occultations of bright x-ray sources by the moon: Shortly after the actual source disappears behind the edge of the moon, the direct radiation is shielded by the moon, while part of the surrounding scattering halo is still visible. At the end of the occultation the sequence is reversed: The halo emerges before the source does. As ROSAT is capable of observing this very faint radiation from the scattering halo, these observations provide direct information about the intensity profile of the scattered radiation, which in turn depends mainly on the size and size distribution of the dust particles.

In this way it was possible to determine the average maximum particle size of the dust in galactic dark clouds, independent from optical measurements but in good agreement with them: They show a size of a ten-thousandth of a millimeter (0.1 μm). A comparison with optical

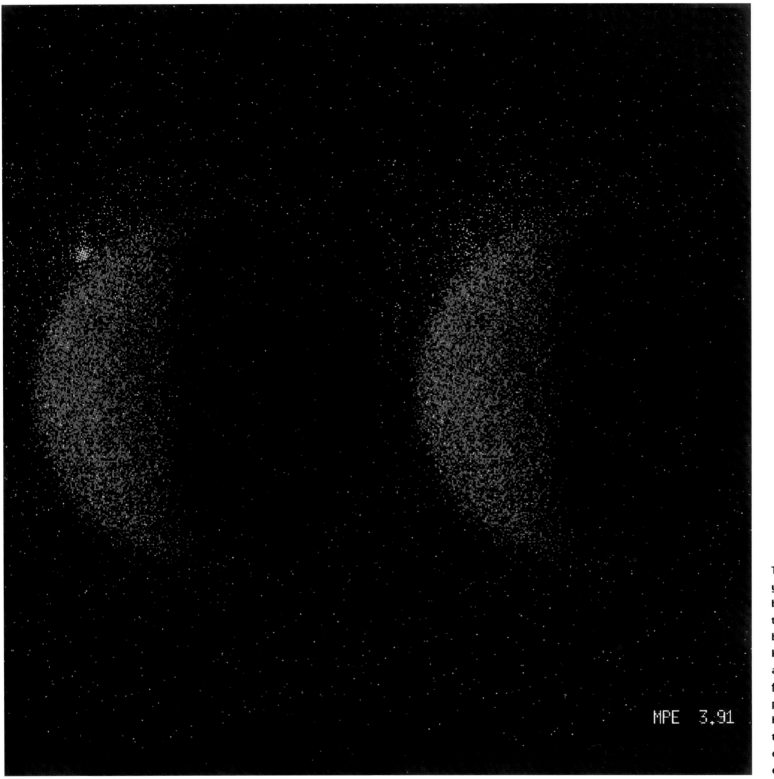

MPE 3.91

The x-ray source GX5-1 (bright yellow dot) is visible shortly before occultation by the moon in the left image; it has disappeared behind the moon on the right. However, the surrounding area still appears x-ray-bright: The radiation from GX5-1 is scattered by dust particles of the interstellar matter between GX5-1 and the moon, so that GX5-1 is surrounded by an extended x-ray halo, which still extends over the rim of the moon.

measurements also showed that the fraction of dust in the interstellar matter is about one percent, the rest being gas.

A Hot Neighborhood

On the one hand, the extinction of soft x-rays in interstellar clouds impedes the study of the invisible sky, but on the other hand it helps us gain information about our vicinity. For some time astronomers have detected diffuse radiation in the x-ray range whose measured intensity distribution indicates that it has to come from hot gas of about a million degrees. But where is this gas? Is it an effect that is confined to the plane of our galaxy (the sun is located only about forty light years above the plane of the Milky Way). Does it surround our entire galaxy like a giant halo, or does this hot gas perhaps fill the entire universe?

To answer this question, the data from the ROSAT sky survey were searched in detail for shadows of interstellar clouds. If the source of radiation is not confined to the immediate vicinity of the sun, these shadows should be noticeable. Indeed, many such shadows have been found, most of them identifiable with known dark clouds. This allows the conclusion that at least part of the diffuse x-ray signal comes from areas beyond the dark clouds, probably from an extended halo, or corona, of hot gas around our galaxy; the diameter of this halo is estimated to be 130,000 light years (compared to our galaxy's diameter of about 100,000).

Strong changes in intensity on small angular scales indicate a fairly uneven distribution of the hot gas. Some of these clumps could be identified as high-velocity clouds, which had been detected by radio astronomers some time ago; these gas clouds move from "above" and "below" towards the plane of our galaxy, with relative velocities of 100 to 200 kilometers per second.

It is easy to imagine that a halo of hot gas has to be continuously reheated. Radio astronomers have found indications for such "heating ducts," which extend from the galactic plane like chimneys, resupplying hot gas to the outside. One of these chimneys originates from the area around a cluster of young, massive stars (Cas OB6) in the southeast corner of the constellation Cassiopeia; the strong stellar winds emanating from these stars have penetrated the surrounding colder gas and are now in the process of advancing into the galactic halo.

Since the diffuse x-ray signal observed by ROSAT does not vanish completely in these shadowed areas, the remaining part (up to fifty percent) must come from the immediate vicinity, from a so-called local hot bubble. Stellar winds from young, massive stars or shock waves of supernova explosions may be responsible for heating this bubble – and therefore for the emission of the soft x-rays.

Bubbles in the Milky Way

The local hot bubble is not the only example of its kind in our galaxy. In 1980, the first "super-bubble" was discovered by the Einstein satellite in the constellation Cygnus (the Cygnus super-bubble), and since then a few more have been found, such as in the region of the constellations Orion and Eridanus (the Eridanus X-Ray Enhancement, or EXE), around the constellations Monoceros

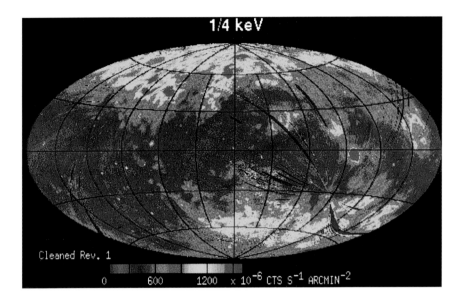

1/4 keV

Cleaned Rev. 1

0   600   1200   x 10⁻⁶ CTS S⁻¹ ARCMIN⁻²

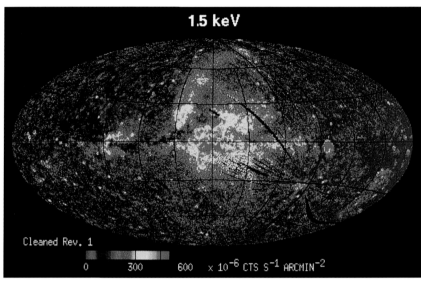

1.5 keV

Cleaned Rev. 1

0   300   600   x 10⁻⁶ CTS S⁻¹ ARCMIN⁻²

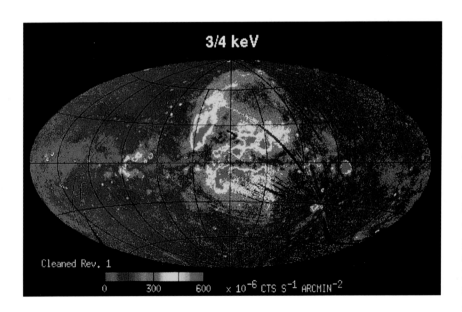

3/4 keV

Cleaned Rev. 1

0   300   600   x 10⁻⁶ CTS S⁻¹ ARCMIN⁻²

This sequence of three images shows the entire sky from the ROSAT survey, with a spatial resolution of two degrees, at three different x-ray energies – 0.25 keV, 0.75 keV, and 1.5 keV, respectively – and in galactic coordinates, with the galactic center in the middle. The low-energy radiation (top left) comes from a surrounding hot gas cloud of about a million degrees; the area of the galactic plane (the darker horizontal strip in the middle) is darkened by interstellar absorption. The relatively close Vela supernova remnant is visible as an isolated red spot. At intermediate x-ray energies (bottom left) we can look at larger distances, and the sky appears more uniform in its intensity but still structured by interstellar absorption. The image is dominated by the radiation from the north polar spur and from the large-scale central region of our galaxy, containing hot interstellar matter and tens of thousands of unresolved point sources. Near the equator in the left half of the image, the Cygnus super-bubble becomes visible as a broken yellow and red ring. At 1.5 keV (right) the sky appears even more uniform, with the exception of the galactic center regions, but the north polar spur has almost faded away.

and Gemini (called the Monogem ring), and in the direction of the constellation Centaurus. The last consists of a giant arc with a diameter of 120 degrees, with Centaurus containing only the geometric center of this huge ring. Its existence was noted by radio astronomers during the 1950s, when they used their antennas to search our galaxy for radiation from interstellar matter; as this region extends from the galactic plane to its north pole, radio astronomers named it the north polar spur. At first, scientists were puzzled by this emission region, which even in the radio sky covers an area of ninety by twenty degrees. With new observations in the radio, optical, and x-ray domains, it is now assumed that this is the remnant of a supernova that exploded between 100,000 and 10 million years ago at a distance of 600 light years.

Comparison of the various energy bands of the ROSAT proportional counters gave the mean temperature at the outer shock front of the expanding gas cloud as approximately three million degrees. After measuring the particle density, the age of the cloud can be estimated to be about 120,000 years.

Similar measurements of the Monogem ring yield an age of about 60,000 years. At a distance of almost a thousand light years, an angular diameter of twenty degrees, and an average temperature of about 1.5 million degrees, it is clearly cooler. Within the ring, though not exactly at its center, lies the pulsar PSR 0656+14, whose distance and age are compatible with those of the supernova remnant. If they are indeed the result of the same event, they would represent the oldest known combination of a pulsar and its corresponding supernova remnant.

Spectroscopy in the optical range shows the Orion–Eridanus super-bubble to be the closest neighbor of the local hot bubble. Their respective shock fronts are about 500 light years away from us and may even begin to cross each other. The Orion–Eridanus super-bubble probably originates mainly from strong stellar winds from the Orion OB1 association of young, hot stars; it has been additionally heated by a supernova explosion some 120,000 years ago. This super-bubble may develop over time into one of the galactic chimneys that contribute matter to the galactic halo.

The Galactic Center

After having discussed the various stellar and diffuse x-ray sources within our galaxy, we should conclude by pointing our x-ray vision at the galaxy's center. This region, located at a distance of 25,000 light years in the direction of the constellation Sagittarius, is off limits to optical astronomers because intervening dust clouds attenuate the light or absorb it completely. As a result, the core of our galaxy remained a blank spot on astronomers' maps for quite some time.

Radio and infrared astronomers were able to "see" through the dust clouds and collect the first images of the galactic center. The density of matter increases steeply towards the center of the galaxy and finally reaches 10,000 solar masses per cubic light year. Radio maps show a tiny minispiral, several light years across, immediately adjacent to the compact source Sagittarius A*, the assumed central object. We find in this region not only

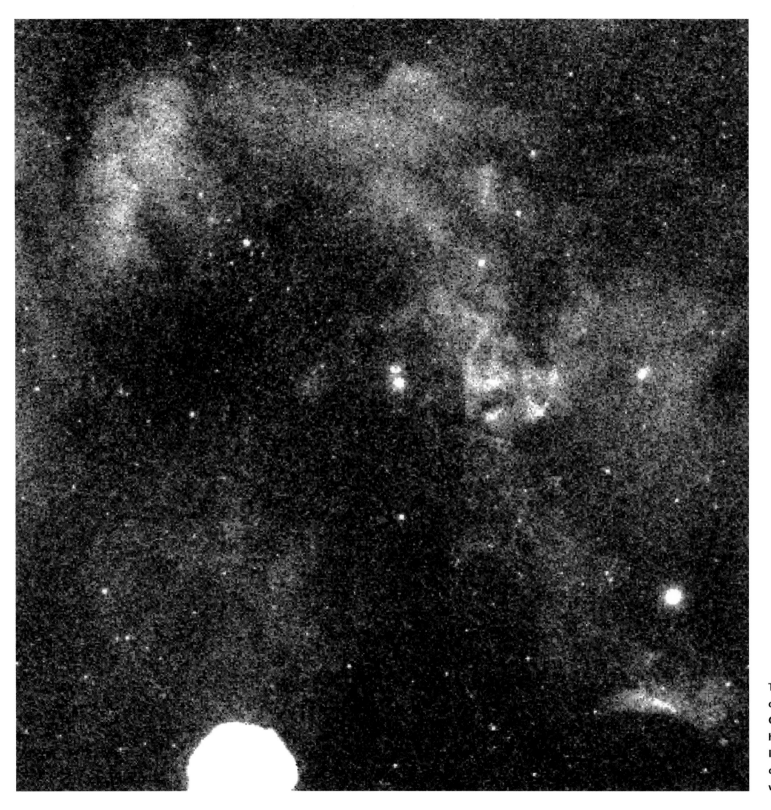

The Cygnus super-bubble is a shell of hot gas in the constellation Cygnus, imaged by ROSAT with high resolution for the first time. It probably originated from a combination of strong stellar winds and supernova explosions.

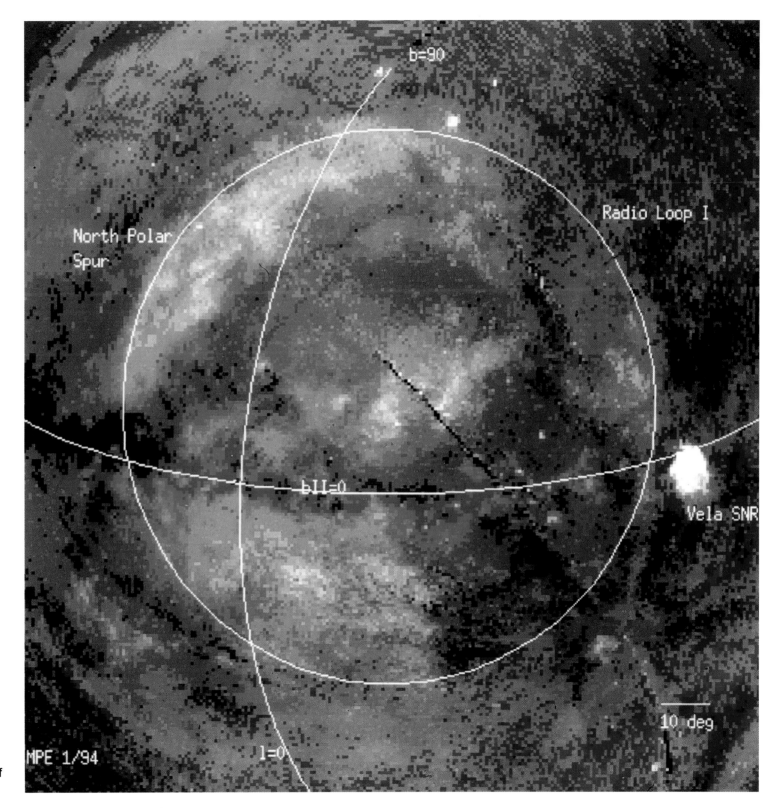

In the x-ray domain, the north polar spur turns out to be the remnant of an old, but close, supernova that exploded perhaps 120,000 years ago at a distance of about 600 light years.

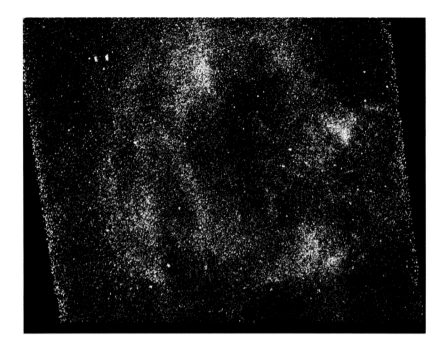

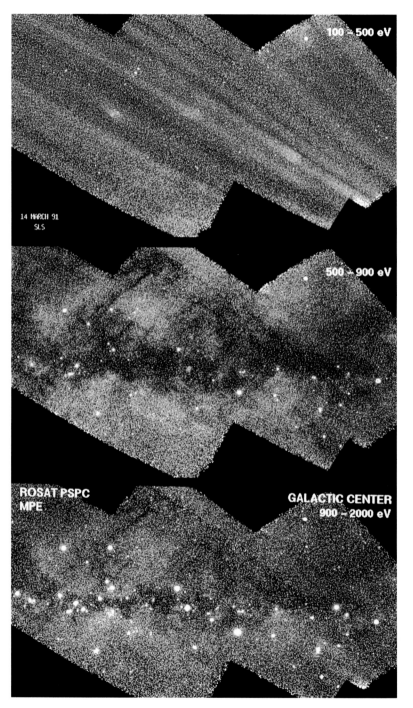

Top: The Monogem ring, a bubble of hot gas of approximately 1.5 million degrees, was created by a supernova explosion about 60,000 years ago.

Right: The view towards the center of our galaxy in the soft x-ray region (top) is inhibited by interstellar absorption, so that only the diffuse radiation from the surrounding hot bubble can be seen. At intermediate energies (middle) sources beyond the local hot bubble become visible. In the bottom image, taken in the high-energy range observable with ROSAT, diffuse x-ray signals from greater distances begin to register. Together with numerous discrete x-ray sources (stars, x-ray binaries, supernova remnants) they are responsible for the apparent structure of the image.

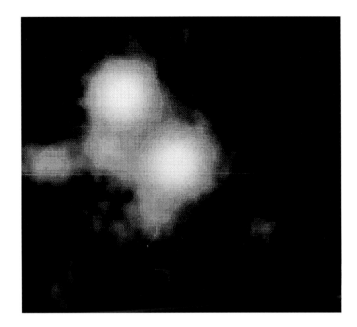 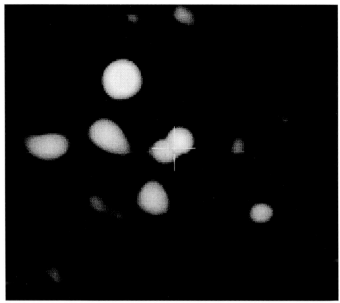

Top: The object in the center of our galaxy can be made visible in the optical range with a trick: By computer, the broadly blurred image of a foreground star was removed from the original data of a 200-minute exposure (left panel). Behind it, two adjacent points of light become visible near the position of the radio source Sagittarius A*, marked by a cross (right panel). The right-hand, brighter object of the two turns out to be a bluish, hot source, which could be the accretion disk of a black hole.

Bottom: The radiation from the galactic center could be resolved into individual sources with ROSAT for the first time: The blue spot at the center of the image coincides almost exactly with the position of the radio source Sagittarius A*, which is identified with the center of our Milky Way. The distinct "blue shift" of this x-ray source indicates strong absorption of soft x-rays, as would exist in the disk of material around a black hole; the red spots are foreground stars, while the radiation corresponding to the diffuse, green areas comes from the central regions of our galaxy.

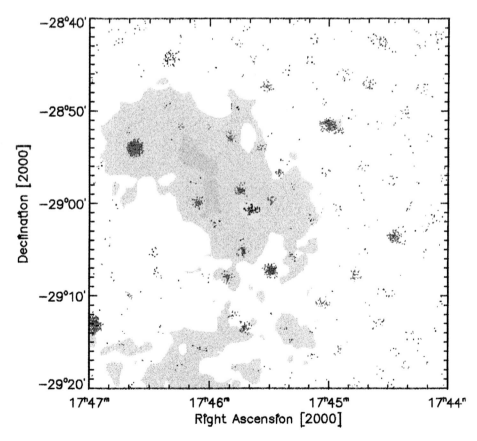

diffuse gas but also several infrared sources, which have recently been identified from high-resolution imaging as dense star clusters. Model calculations indicate that the density of stars in the immediate vicinity of the galactic center is great enough to enable stars to collide and even merge.

Perhaps a Black Hole?

Scientists who have tried to calculate the motion of objects within the core regions of our galaxy from radio and infrared spectrograms have determined that a black hole of about one million solar masses has to exist in the center. However, this black hole does not appear to be very "active," since the radiation from its immediate surroundings reaches just one million solar luminosities. Active black holes of this mass easily reach up to thirty billion solar luminosities; this lack of activity must be related to the very low rate of infalling matter – at most several millionths of a solar mass per year.

The Einstein satellite had already identified an x-ray source in the direction of the galactic center. Additional observations carried out by the Spacelab-2 mission in 1985 showed several distinct hard x-ray sources in this area, which have since been confirmed by ROSAT at lower energies. It turned out that the source RXJ 1756.6-2900, which coincides with the position of Sagittarius A* to within ten arc minutes, experiences much stronger absorption than any of the other sources in this direction. This could mean that it is simply a faraway background object. But it could also mean that it is a much stronger source, heavily obscured by a surrounding accretion disk and stripped of most of its soft x-ray signal.

Did ROSAT find the black hole in the center of our galaxy after all?

X-Ray Astronomy Outside Our Galaxy

At the beginning of the twentieth century, most astronomers believed that the Milky Way was the only object of its kind and represented a large fraction of the universe. With increasingly larger telescopes (the Yerkes refractor with an objective 41 inches in diameter had begun operation in 1897), they had observed faint distant "nebulae," but they did not succeed in identifying their true nature at that time, nor in measuring their distance. Only a few had speculated that they were distant islands of stars similar to our galaxy, as Immanuel Kant had suspected already as early as 1755.

Since then, our understanding of the universe has changed dramatically. After Edwin Hubble, using the 100-inch reflector at the Mount Wilson Observatory, photographed individual stars in the Andromeda galaxy and even determined its distance during the 1920s, everything fell into perspective. Today, astronomers describe the universe in which we live (at least its observable part) as a vast realm containing many billions of galaxies, extending more than twelve to fifteen billion light years in any direction and continuing to grow.

Neighboring galaxies offer optical astronomers the opportunity to complete their understanding of our own galaxy. Because light from the stars is weakened by the dark clouds in our galaxy, only a small fraction of the Milky Way can be observed and explored in visible light.

This puts astronomers in a position like that of a visitor to a foreign city, trying to estimate its size or even to generate a rough map of the main streets based on the view from the hotel room. This may be possible for the immediate vicinity, as high trees or street lights indicate roads or church towers indicate the existence of more buildings. But somewhere these indicators fade into a sea of houses. The size and structure of the city would be much clearer if one could fly over it.

The situation is similar for other wavelengths, from the radio domain to x- and gamma-rays. Familiar objects may be found in other galaxies, providing, for example, clues about their potential role in the development of different types of galaxies. On the other hand, completely new circumstances may be found whose exploration may lead to unexpected links and connections. Finally, astronomical exploration of galaxies and of the universe as a whole suffers from the same restriction as that encountered in the study of stars: Scientists cannot actively influence events or carry out experiments, but are confined to observing as many objects as possible. As expected, ROSAT's extragalactic observations provided many surprises.

Our Nearest Neighbors

Tourists considering themselves connoisseurs may rave about palmy beaches under the Southern Cross, but this constellation is less than impressive, and many who pretend to have basked in its light probably looked for it in vain. Cognoscenti of the southern skies would rather look for the two Magellanic Clouds, which allow them–at least in their thoughts–to traverse great distances: While the two Magellanic Clouds appear to the naked-eye observer as two separated parts

The Large Magellanic Cloud in the Southern sky is a small neighboring galaxy of our own Milky Way, at a distance of about 170,000 light years.

of the Milky Way–as bright clouds of stars located more than thirty degrees from the plane of the Milky Way–they are actually more than 170,000 light years away from us. This makes them small, independent galaxies, circling our own like satellites.

Their distance, small by astronomical standards (the closest larger galaxy, the Andromeda Nebula, is more than two million light years away), makes the Magellanic Clouds rewarding targets for astronomers. On the one hand, they are near enough so that individual stars of a hundred times solar luminosity can be observed with fairly modest telescopes. On the other hand, as largely independent stellar systems, the Magel-

lanic Clouds have undergone a different evolution from the Milky Way's in spite–or maybe because–of their closeness to our much larger galaxy. With 10 and 2.5 billion solar masses respectively, the two Magellanic Clouds are significantly smaller than our own galaxy, whose mass is estimated from many different observations to be about 100 billion solar masses.

During a sounding rocket flight in 1968, x-rays from the direction of the Large Magellanic Cloud (LMC) were registered for the first time; but because the signal hardly exceeded the background noise, not much information could be gained. During another rocket flight two years later, several bright x-ray sources were detected

LMC
ROSAT PSPC
SURVEY
color image

2 degree

This x-ray true-color image of the Large Magellanic Cloud, generated after completion of the ROSAT sky survey, shows more than 500 new sources. While foreground stars and supersoft sources, found for the first time, are shown in red, and the diffuse gas of about five million degrees in green, x-ray binaries, supernova remnants, and far more distant active galaxies and galaxy clusters are conspicuous in hard x-rays, shown in blue.

in the LMC and one more in the Small Magellanic Cloud (SMC) with a strikingly hard spectrum.

The x-ray satellite Uhuru, launched into orbit a short time later, was not able to contribute much more, but the Einstein satellite increased the number of known sources to more than one hundred from the LMC and almost fifty for the SMC. Thanks to improved spatial resolution, many of these x-ray sources could be matched with optical counterparts. Again, the brightest sources turned out to be x-ray binaries, followed by supernova remnants. At least one of these expanding gas clouds closely resembles the Crab Nebula–even down to a fifty-millisecond pulsar, whose pulses can be seen in both the x-ray and optical domains.

One of the greatest surprises was the discovery of very bright sources in the LMC; they release enormous amounts of energy, similar to bright x-ray binaries. But their spectra indicate a temperature of only several hundred thousand degrees, unlike the several million degrees typical for x-ray binaries. Some of these sources had been noticed by the Einstein satellite because of their soft spectra, but its spectral resolution was insufficient to determine their temperatures. This led to speculations about soft x-rays from black holes.

ROSAT measurements showed that these sources typically had diameters of about 10,000 kilometers. That at first led scientists to believe they were seeing cocoons, or gas envelopes, around accreting neutron stars, as had been suggested for low-mass x-ray binaries. But after more and more of these sources were discovered, in the Small Magellanic Cloud and Andromeda galaxy as well as the LMC, this theory could not be sustained. Why should all these cocoons have exactly the size of a white dwarf?

The scientists were finally led to a model that theoreticians had developed earlier in a different context. It involves a double star system in which a white dwarf obtains just enough mass from its partner (about a third of a lunar mass per year) for continuous fusion processes on its surface. In principle, these supersoft sources are a special case of cataclysmic variables.

However, this model demands a careful balance, because if not enough mass is supplied, the fusion motor will begin to stutter, and with too much fuel the fusion rate will increase, quickly producing an extended atmosphere around the white dwarf and giving it the appearance of a red giant. Although ROSAT has found several dozen sources of this kind in the Magellanic Clouds and the Andromeda galaxy, they are much harder to detect in our own galaxy because absorption by interstellar matter diminishes the predominantly long-wavelength x-rays from these sources.

A New X-Ray Pulsar

The ROSAT sky survey detected a total of 516 x-ray sources in a 13.5 by 13.5 degree area of the sky centered on the LMC, four-fifths of which were previously unknown. Although this survey included foreground stars as well as distant galaxies, nearby objects can normally be excluded based on their identification with known radio, infrared, or optical sources, or by applying interstellar and intergalactic absorption as a yardstick. There is relatively little absorption of soft x-rays from foreground stars, while this absorption for more distant galaxies is higher than the value for the LMC.

The 516 sources the scientists from Garching found included the first pulsar discovered solely from x-ray observations. This source had been seen by EXOSAT already during the eighties, but the length of the observations was insufficient to detect a pulsed x-ray component. ROSAT's forty observing sequences, obtained over a period of four years, suggested that the source was strongly variable. When the scientists analyzed four of these sequences, they discovered x-ray signals with a period of 13.67 seconds.

From the spectrum of the optical counterpart it is known that this pulsar moves around a massive Be star and apparently collects matter from its stellar wind. (The suffix "e" denotes the presence of emission lines in this B-type star, indicating surrounding luminous gas clouds.) Since the pulsar alternates between moving towards us and away from us in its orbit, the periodic shift of the arrival times of the pulses can in principle be detected in long-term continuous observing runs, leading to the direct derivation of the orbital period. Optical astronomers had already learned to extract such results from intermittent data, as their carefully planned long-term observations had often been interrupted by bad weather. Using the same method, the ROSAT measurements yielded a most probable orbital period of 25.4 days and a mean distance of seventy million kilometers, approaching three times the stellar diameter.

Super-Bubbles in the LMC

In our own galaxy we can only observe closer super-bubbles (the soft x-rays from more distant super-bubbles are too strongly absorbed), but the two Magellanic Clouds offer the possibility of a complete survey.

Like a light in a fog, the diffuse x-ray signal of a hot gas cloud is difficult to distinguish from the confusing background by its shape alone; therefore, it is necessary to suppress this background as much as possible. The Einstein satellite was able to distinguish the brightest sources in this fog only after prolonged probing,

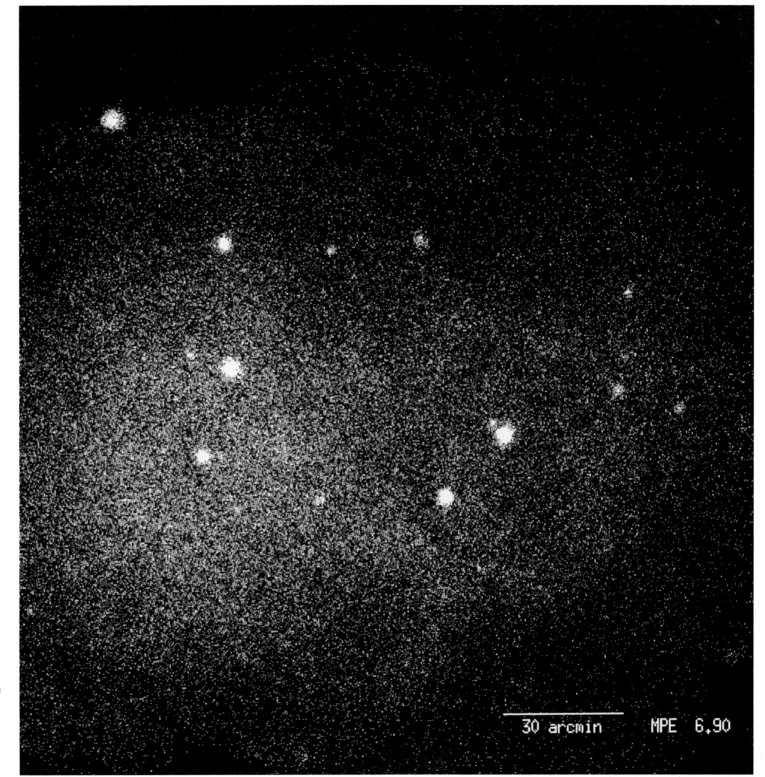

The Large Magellanic Cloud exhibits extended diffuse radiation in the low-energy x-ray range; it comes from hot gas between the stars. Several supersoft sources and foreground stars are also visible.

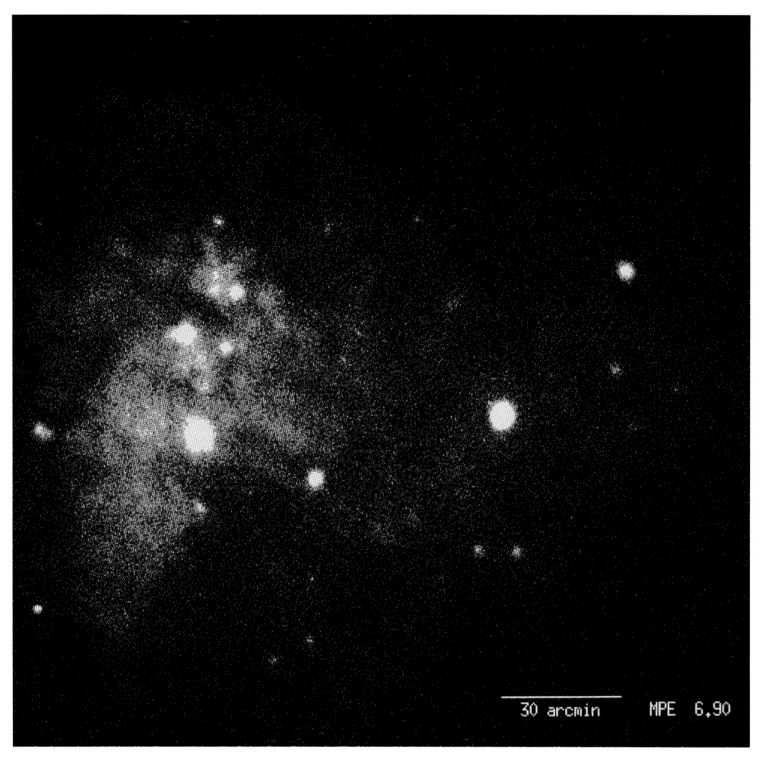

30 arcmin   MPE 6.90

At higher ROSAT energy ranges, the source LMC-X1 left of image center becomes the brightest object. Above this source and fairly close to each other, a foreground star (right) becomes visible, together with a very bright supernova remnant, surrounding a fifty millisecond pulsar, and the Tarantula Nebula, with the conspicuous shadow of a dark cloud.

but ROSAT, with its better-shielded detectors, showed the distribution of the hot gas more or less immediately, and a detailed examination of individual structures, such as star-forming regions or hot super-bubbles, now became possible. One of the questions about hot super-bubbles concerns the origin of the observed x-ray intensity.

The standard model postulates that expanding fast stellar winds produce a growing cavity in the surrounding interstellar matter; as they are suddenly slowed down just inside the boundary layer of the cavity, they are sufficiently heated to generate the observed x-ray signature. But in some cases the x-ray intensity is higher than this model predicts.

Currently, three possible enhancement mechanisms are being discussed. First, the surrounding interstellar matter may not be of uniform density, so that in some places within the boundary layer more hot gas is available to emit x-rays. Second, the shock front may have been heated in addition by close supernova explosions. Or third, clouds of matter within the super-bubble, heated by nearby supernova explosions, increase its "inner" x-ray intensity. Since not all super-bubbles in the LMC are truly brighter than expected from the standard model, the increase in intensity may well be temporary, which favors the enhancements by supernova explosions.

For the super-bubble in the LMC that shines brightest in the x-ray spectrum (N44), this connection may be applicable. Not only is a supernova remnant visible in the identical direction, but its x-ray intensity is comparable to that of N44. But the "normal" x-ray intensity of four other super-bubbles can be explained with the standard model alone, without supernova enhancement.

The 30 Doradus Complex

Unlike supernova remnants and x-ray binaries, the star-forming regions in the LMC show a distinct concentration toward its eastern part, which faces our own galaxy. This area contains the 30 Doradus complex, which is visible with the naked eye; because of its distinctive shape, it is also called the Tarantula Nebula. It consists of a huge gas cloud of about half a million solar masses. In its center, a large cluster of very massive and extremely hot young stars has been confirmed; their energetic ultraviolet radiation ionizes much of the surrounding gas up to a distance of 350 light years, inducing it to glow. Extended gas clouds that are excited to emit light by neighboring or enclosed hot stars are called H-II regions. The Tarantula Nebula is the largest known H-II region and is much larger than comparable star-forming regions in our own galaxy. The enclosed hot stars also produce x-rays, but they can only contribute a few percent of the observed x-ray intensity of this region; an additional ten percent comes from a known supernova remnant (SNR 30 Dor B).

As part of a dedicated observation in a field of twenty by twenty arc minutes (about half the area of the full moon) centered on 30 Doradus, ROSAT has seen a total of 34 x-ray sources: foreground stars, known supernova remnants, and other sources. From the energy distribution of the soft radiation in the central part of the Tarantula Nebula, the temperature of the diffuse

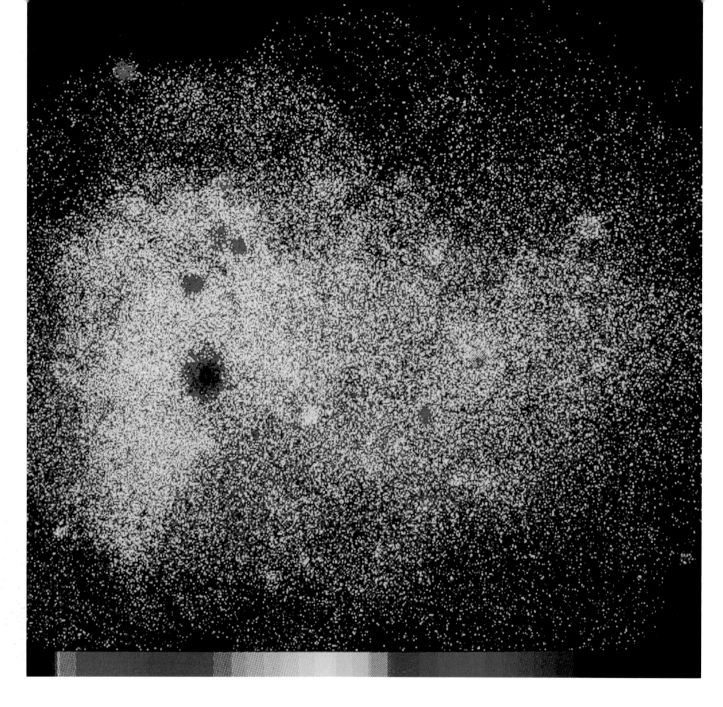

gas was estimated at three million to five million degrees.

The Andromeda Galaxy

On a clear, moonless fall night, at dark sites away from large cities, a faint nebulous spot can be seen with the naked eye high in the sky. Larger amateur

This color-coded x-ray image of the Large Magellanic Cloud clarifies the temperature distribution of the interstellar gas: In the red areas the temperature is around 500,000 degrees, in the blue supernova remnants and super-bubbles it is about 10 million degrees.

The Andromeda galaxy M 31 is our closest neighboring large galaxy, a spiral galaxy at a distance of about 2.25 million light years. The green crosses mark the positions of x-ray sources detected by ROSAT.

telescopes show an oval cloud, revealing some details (dark clouds) in longer exposures. With a bit of effort one may imagine a spiral structure, foreshortened by perspective. The eighteenth-century French comet hunter Charles Messier, compiling a small catalog of nebulosities in the sky to avoid being distracted by "false comets," entered the Andromeda Nebula as number 31 (M 31) on his list of a hundred-plus entries.

Only in the last seventy years has the An-dromeda Nebula been recognized as a large neighboring galaxy. Not until its distance had been determined at 2.25 million light years did we know that M 31 is a big brother of the Milky Way with a diameter of 160,000 light years and a total mass equal to that of 320 billion suns.

The first x-rays from Andromeda's direction were registered during a rocket flight in February 1973. In the same decade, the Einstein satellite observed about thirty bright sources in the outer regions of the spiral arms of the Andromeda galaxy, as well as a central "spot" representing numerous individual sources; the increased resolution of the HRI detector (an improved version of which is now deployed on ROSAT) was able to discern many more individual sources in the central cluster and increased the total number of observed objects to 108.

A relatively bright object, only five arc minutes from the center of M 31, attracted attention because its brightness changed by a factor of ten during the observing period of six months. This behavior is characteristic of so-called active galactic nuclei, but the x-ray intensity of this central source in the Andromeda galaxy falls short of active galactic nuclei by a factor of several million – like the central object in our own galaxy. This could support the hypothesis that all galaxies contain such active nuclei, but many have been "starved" almost to extinction and emit only a minimum of energy.

From the outside, our galaxy probably looks a lot like M 31: a bright nucleus with elliptical cross-section (the central bulge), surrounded by a flat disk with embedded spiral arms. From our location

20 arcmin   MPE 8.91

A mosaic of the Andromeda galaxy
M 31; most of the more than 500
sources are x-ray binaries and
supernova remnants.

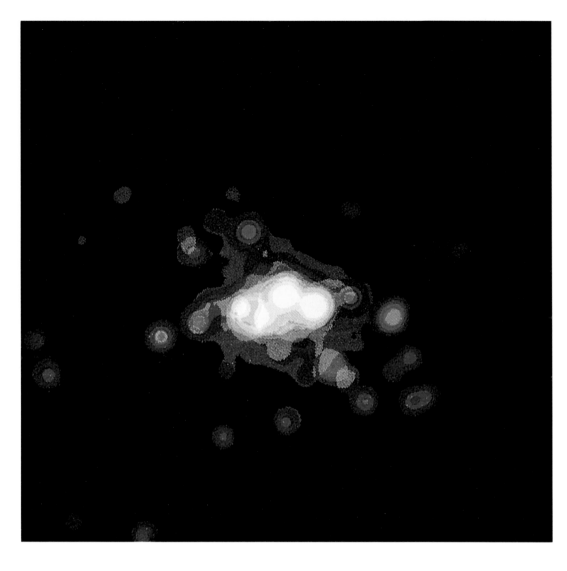

The central region of the Andromeda galaxy contains a collection of hard x-ray sources (blue), which are x-ray binaries. The red dot is a foreground star, while the yellow one corresponds to a supersoft x-ray source in M 31. The diffuse component of radiation is due either to fainter, unresolved point sources or to hot gas of several million degrees.

within the galaxy, this disk appears as the shining band of the Milky Way in the sky.

The interest of scientists in observing x-ray sources in the Andromeda galaxy and comparing their distribution, brightness, and characteristics with those of galactic sources was understandably large.

ROSAT confirmed about 560 sources in the area of the Andromeda galaxy – more than Uhuru found in the entire sky during the seventies – all of the same types as those identified by Uhuru in our own galaxy: x-ray binaries, supernova

remnants, and a few extragalactic sources. Optical and radio counterparts have been found for just over 20 percent of them. Among them are 55 foreground stars from our own galaxy, 33 sources in globular clusters, 23 supernova remnants in the Andromeda galaxy, and 10 extragalactic background sources.

Statistical analysis combined with absorption measurements permit us to estimate the types of the remaining unidentified sources. Of the 560 sources, 370 are located in the Andromeda galaxy itself. From their energy distribution it can be deduced that they are mainly x-ray binaries, but a few transient and several supersoft sources were also found.

Low-mass x-ray binaries appear to dominate in the region of the central bulge. The thought that they could be late evolutionary stages of "burnt-out" novae had to be abandoned, however, when it turned out that these sources had a significantly different spatial distribution than newly occurring novae. Instead, there seemed to be a link between the distribution of the low-mass x-ray sources and the inner bases of the spiral arms, indicated by clouds of ionized hydrogen (H-II regions). This link is not necessarily compelling on theoretical grounds, but it may tell us something about the formation of x-ray sources in the central bulge.

If all point sources are removed from the ROSAT data, only the diffuse x-ray signal should remain. When we do this we indeed find diffuse radiation in the central part of the Andromeda galaxy; further out, in the zone of the spiral arms, x-rays may be generated as well, but there they are absorbed by the considerably denser interstellar matter. But it is unclear whether this radiation comes from numerous faint individual sources that cannot be resolved in the images or if it is a diffuse gas component. If hot gas were solely responsible for the observed radiation, it would have to amount to 1.3 million solar masses.

Starburst Galaxies

While M 31 and our own galaxy are comparatively quiet, infrared and radio observations put NGC 253, a galaxy in the constellation Sculptor, into the group of so-called starburst galaxies. In the infrared, its core region appears as one of the brightest sources, despite its distance of about ten million light years, and radio astronomers observe indications of a fast stream of matter emanating from the core region in both directions perpendicular to the galactic plane. These observations together lead astronomers to believe that the core region experienced a phase of violent star formation (a "starburst") relatively recently. Such an event produces a large number of massive stars with correspondingly fast development cycles, so that they soon explode as supernovae, producing a large quantity of hot, x-ray-emitting gas that finally flows out of the galaxy.

The Einstein satellite had observed x-ray emission in NGC 253 in several locations. An extended source apparently exists in the core region, and bright point sources coincide with the spiral arms. In addition, Einstein had found indications for diffuse radiation from the core toward the south, in the inner region of the disk, and from the northern half of the halo. With ROSAT it was possible for the first time to observe the diffuse x-ray emission of a hot gas of two million degrees out to a distance of 30,000 light years in the halo of the galaxy, on both sides of the disk. The measured x-ray intensity tells us the surprisingly low density of the hot gas: only a little more than fifty atoms per cubic foot. But when this number is summed over the huge volume of the galactic halo, we obtain a total of at least six million solar masses. If this gas was ejected from the core region as the result of starburst activity, this event must have occurred tens of millions of years ago – this assumption agrees well with the results from other analyses. The conspicuous absorption of the diffuse soft halo radiation by an intervening spiral arm is also clearly observable.

For the halo to be permanently bound by the galaxy's gravitation so that it does not expand into intergalactic space, NGC 253 must contain about fifty billion solar masses. Alternatively, the so-called rotation curve of the galaxy can be determined by using spectroscopy to measure the rotational velocity as a function of distance from the center; applying the laws of celestial mechanics yields the total mass necessary to make the observed rotation possible – about fifty billion solar masses as well. At first glance this may be reassuring, but this "calculated" mass is much larger than can be accounted for by the stars in that galaxy. The difference can be explained only by assuming some form of dark matter, which apparently does not emit electromagnetic radiation but only makes itself felt by its gravitational force. We will encounter this phenomenon a few more times.

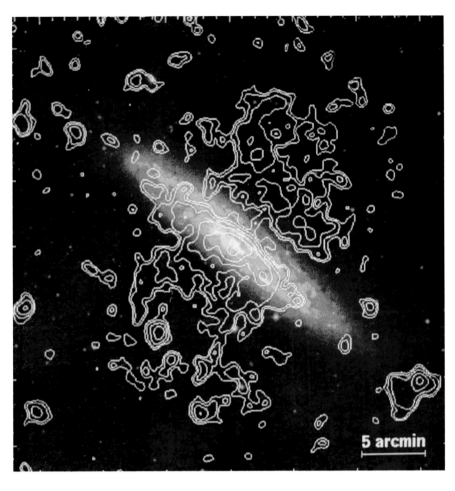

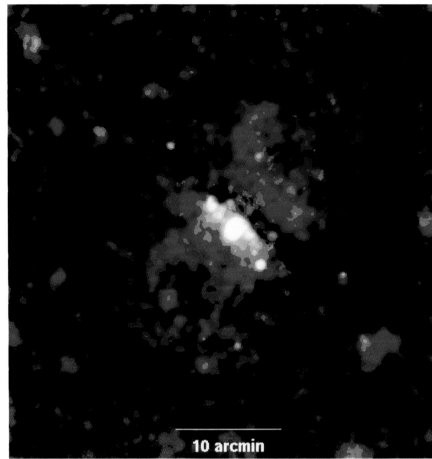

Left: In the starburst galaxy NGC 253, ROSAT discovered radiation from hot halo gas of about two million degrees. The contours of x-ray intensity are overlaid on an optical image.

Right: The hot halo gas surrounding the core of the starburst galaxy NGC 253, strongly visible in the upper right and lower left areas, has a softer x-ray spectrum (red) than the core of the galaxy (blue–white).

The outflow of matter and the corresponding x-ray emission are most easily observed in edge-on galaxies, where we are looking at the galaxy from the side. Examples of such galaxies are NGC 4565 (in the constellation Coma) and NGC 4656 (in the constellation Canes Venatici). In addition to point sources in their core regions and disks, ROSAT found diffuse emission centered on the core or on a region disturbed by the tidal forces of a close neighboring galaxy and extending far outside the disks.

X-ray halos were also observed in the galaxies NGC 4631 (in the constellation Canes Venatici) and NGC 3079 (in Ursa Major). NGC 3079 also shows a central source with a diameter of about twenty arc seconds, which coincides with a super-

bubble emerging from the core of the galaxy. Such a super-bubble was thought to exist in NGC 4631 as well, but the HRI detector showed that it was actually the superposition of several point sources, probably x-ray binaries.

Canes Venatici also contains the perhaps most impressive spiral galaxy: M 51, the prototype of all spiral galaxies. The English astronomer William Parsons, the third Earl of Rosse, first saw a spiral structure in this nebula in 1845, capturing it in a drawing of M 51. This galaxy, also known as the Whirlpool galaxy, is circled by a smaller system (NGC 5195), giving rise to strong interactions with significant influence on their inner structures. A stream of matter probably flows between the two galaxies. In a ROSAT image NGC 5195 appears

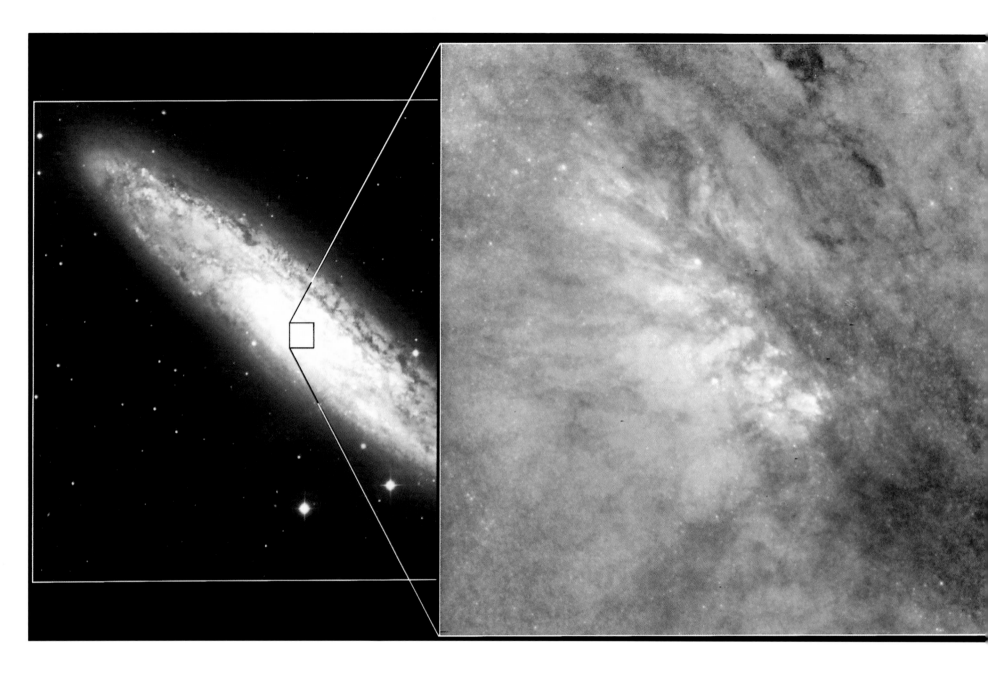

The starburst region in the center of the galaxy NGC 253 – in this high-resolution image from the Hubble Space Telescope – has a diameter of about one thousand light years.

NGC 4565
ROSAT PSPC
0.5–2.0 keV

2 arcmin

more easily dissipate into the halo. But it is also possible that so-called magneto-hydrodynamic waves, channeled by these magnetic structures, heat the halo gas to the temperature required for x-ray emission.

For several hundred million years, the two adjacent galaxies NGC 4038 and 4039, in the constellation Corvus, have been involved in a cosmic dance around each other, the results of which are noticeable in all wavelengths. In visible light two long extensions are apparent, reminiscent of an insect's antennae and leading to the name Antenna galaxy; they are the result of strong tidal forces that have pulled large amounts of gas and stars from the two galaxies. In connection with this effect, avalanches of star formation occurred, leading to the observed bright infrared intensity. ROSAT data show thermal radiation indicating hot gas of several million degrees, in addition to the spectral distribution typical of x-ray binaries.

Active Galaxies

The farther we move from our galaxy, the more the single sources take a back seat in ROSAT images. Their x-ray intensities are typically insufficient over large distances to cause a clearly recognizable signal in the detector within practical observing times, and in any case its limited resolution prevents the identification of small details. For more distant galaxies, observers concentrate on large-scale structures and processes in the core that supply the energy for radiation, including x-rays.

An x-ray halo around the core of a galaxy can be seen particularly well when the galaxy is viewed edge-on, as with the spiral galaxy NGC 4565. The optical image (color) is overlaid with the contours of x-ray intensity.

connected to M 51 by a bridge.

There are clear correspondences between the HRI image of this pair of galaxies and their thermal radio emission. Both are connected with star formation.

Interestingly enough, several areas of very diffuse x-ray emission in M 51 (and also in M 83, only ten million light years away) coincide with regions where the radio emission exhibits strong magnetic influences. This could be an indication of magnetic deviations leading from the galactic disk to the halo region. In such regions, which may be related to the galactic "chimneys" known from our own galaxy, the hot x-ray-active gas could

A relatively close example, at a distance of fifteen million light years, is the galaxy NGC 5128, which first attracted the attention of radio astronomers during the 1960s. As Centaurus A, the brightest radio source in the constellation Centaurus, this object had been registered by the 1940s, but the increased resolution of larger antenna dishes was needed to reveal the gigantic size of this source. In the radio sky, the emission from Cen A covers an angle of more than ten degrees, corresponding to a length of three million light years.

In contrast, the optical counterpart, an elliptical galaxy with a diameter of about 100,000 light years, appears quite tiny. It not only shows a pronounced dust lane around its equator but also attracts attention with its immense luminosity, corresponding to about twenty million suns. Thus arose the early suspicion that NGC 5128 might be the result of a collision in which an elliptical galaxy had swallowed a smaller spiral galaxy.

This hypothesis, fairly daring for its time, was rejected then but has recently become more attractive now that astronomers have found indications of similar collisions elsewhere in the universe. Such galactic collisions–and also close encounters–are considered important events in the development of galaxies; in the earlier, much smaller universe, they also must have been much more frequent than today.

From high-resolution radio observations we know that thin jets of matter supply energy to the gigantic areas of radiation far outside the actual galaxy. Nodule-like concentrations within these jets indicate that larger pulses of energy

are delivered only occasionally. The spectrum of the observed radio emission, together with measurements of polarization, reveals that this energy is synchrotron radiation.

The Einstein satellite also showed several nodules in the x-ray domain, in many cases coinciding with the location of radio-bright nodules. The increased spatial resolution of ROSAT now showed that these nodules have diameters of about 150 light years. The partial absorption by the intervening dust lane is not sufficient to mimic nodule-like structure in the x-ray jets.

Interestingly enough, the diffuse x-ray emission continues in the southern extension and reaches again a maximum at its end, while the northern extension appears to be x-ray dark. Together with radio-astronomical observations this

The galaxy NGC 5128 exhibits an unusual dust belt in visible light. Radio astronomers know it as Centaurus A, the closest radio galaxy, with a distance of ten to fifteen million light years. (Source: ESO.)

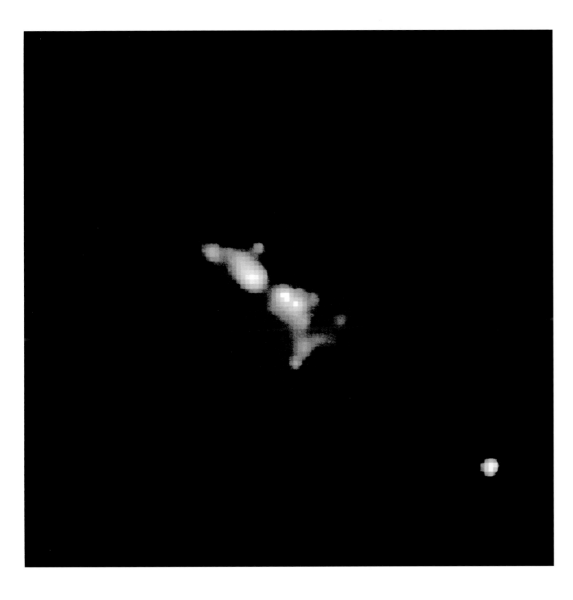

Close to the center of NGC 5128 (Centaurus A), ROSAT observed a jet, which apparently supplies the extended radio regions further out with energy.

gives the impression that in the southern extension we are looking at the front of the jet as it plows into the surrounding intergalactic matter, observing the soft x-ray emission generated there; in the northern extension this active region would be on the far side, so that its radiation would be heavily absorbed by intervening matter and remain undetectable.

There is also a jet in the giant elliptical galaxy M 87, located about seventy million light years away in the center of a large cluster of galaxies (see page 156). This jet was discovered by the American astronomer Heber Doust Curtis in 1918, but it remained an incomprehensible curiosity for a long time. During the 1950s, M 87 turned out to be a strong radio source, and soon it was shown that most of its radio emission came from the area of the jet. Optical and radio observations in the mid-1950s showed that this was synchrotron radiation, which meant that the jet had to consist of charged particles moving in spiral tracks along magnetic field lines. About ten years later, comparatively strong x-ray emission from this region was found, leading to M 87 being recognized as the first extragalactic x-ray source.

While the jet of M 87 is only about 6500 light years long–considered extremely short compared to others in extended radio galaxies–it still poses the problem of resupplying energy. Even highly energetic electrons can emit x-rays for only a limited time, and as they lose energy their emission has increasingly longer wavelengths. They could hardly travel several thousand light years and still emit x-ray photons.

In the eighties, observations with the Very Large Array in New Mexico, a radio telescope consisting of 27 single parabolic antenna dishes, revealed nodule-like intensity peaks in M 87's jet with much darker spaces between them; this argued for the electrons being reaccelerated on their way out. The interesting question remained whether this resupply of energy was large enough in all cases to stimulate x-ray emission.

This question was answered by ROSAT observations with the high-resolution HRI detector. There appear to be only two zones where the x-ray emission comes from synchrotron radiation; other areas contribute little or nothing to the total x-ray emission of the jet.

From other measurements, we know that M 87 is a monster of a galaxy. Its diameter is estimated at more than 500,000 light years, and it may contain several trillion stars. In the late 1970s astronomers attempted to derive the mass of the central core from the movement of stars within this galaxy. They arrived at a value of about five billion solar masses but could not find a correspondingly bright central region – apparently, most of the matter remains dark. Several astronomers suggested that M 87 might harbor a huge black hole, sucking up matter from its vicinity at irregular intervals and thus releasing energy to supply the jet.

Since then, the Hubble Space Telescope has elevated this conjecture to near-certainty. From its vantage point above the perturbing atmosphere of Earth, much more accurate measurements are possible than from the ground. These observations virtually exclude all other explanations for the central object.

An End to the Confusing Variety?

Compared with other galaxies with active core regions, called AGNs (active galactic nuclei) for short, M 87 has to be considered fairly harmless. The broad term AGN combines all somewhat unusual galaxies with strong core activity and short-term variability. These range from radio galaxies, observed at first in the long-wavelength region of the electromagnetic spectrum; to Seyfert galaxies, whose special characteristic – an extremely bright core region – was first noted by the American astronomer Carl Seyfert during the 1940s; to BL Lacertae objects, whose spectra are largely devoid of any structure; to quasars, which were discovered by radio astronomers. After more than thirty years of intense research, a common scenario has begun to emerge. It appears that in all these classes of objects – initially seen as distinctly different – a similar "central engine" is at work, and that the variety of appearances comes from our seeing them at different angles, or from their differing amounts of supplied matter.

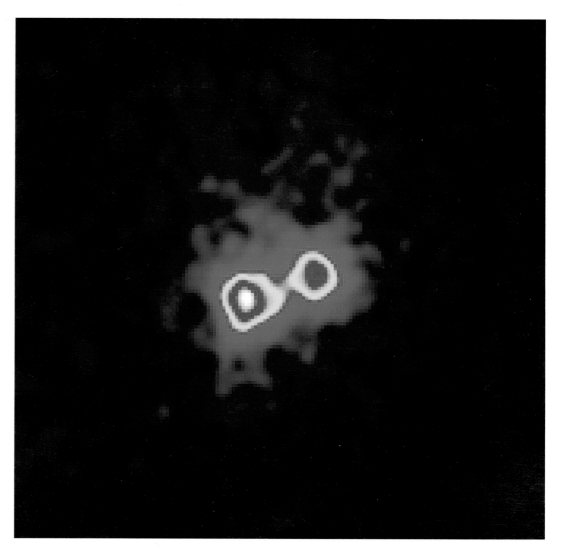

ROSAT also found x-rays from a jet in M 87, probably originating from a black hole in the center of the galaxy. M 87 is a giant elliptical galaxy in the center of the Virgo cluster, at a distance of about seventy million light years.

The radio galaxy NGC 1275 in the Perseus cluster, at a distance of about 300 million light years, contains an active core, conspicuous both in the radio range (contours) and in the x-ray regime (color image). The diffuse x-ray component comes from hot gas of about fifty million degrees between the galaxies, which apparently gets pushed away in areas of high radio radiation.

According to this scenario, a massive black hole, with its extreme gravitational field, will always release a huge amount of energy. It sits like a parasite in the center of a host galaxy and swallows matter from its immediate surroundings. Also, on this galactic scale accretion disks develop in which matter moves inward in increasingly tight spirals, being heated to extreme temperatures; and very strong magnetic fields can again cause a fraction of this matter to be ejected in the two directions perpendicular to the disk, producing a jet structure.

If we look at such a system from the side, the surrounding galaxy blocks the view of the accretion disk, and we see only the jets extending into the galactic halo, where they generate large areas of radio emission. Examples of such radio galaxies, the innermost parts of which are also frequently identified as x-ray sources, are Centaurus A and Cygnus A. However, if we look directly into such a jet, the effect is like staring directly into a headlight: The jet appears extremely bright. This variant corresponds to the class of BL Lacertae objects.

At other angles of view, we may see partial absorption of the emitted radiation in clouds of matter above the inner rim of the accretion disk, or scattering on extremely hot gas in the vicinity of the black hole. These effects cause the observed spectral characteristics of Seyfert galaxies. BL Lacertae objects and Seyfert galaxies differ in another way as well. While BL Lacertae objects are associated with elliptical galaxies, Seyfert galaxies show a spiral structure.

Finally, quasars, mostly observed at much larger distances and therefore at much earlier stages of their development, can be understood as massive black holes that are generously supplied with matter (and therefore energy) by young host galaxies.

More and more astronomers consider it possible that every galaxy has a massive black hole at its center. Once these objects have swallowed all the matter in their vicinity, they fall into a kind of cosmic hibernation, from which they can only wake when new matter is supplied from the outside, for instance by a close encounter or even a collision of their host with another galaxy. This strongly disturbs the distribution of matter in both galaxies, so that large quantities of matter can fall toward the center again.

Multispectral Cooperation

While active galactic nuclei are fairly inconspicuous in the optical domain and difficult to find in the sea of foreground stars, they take center stage in the x-ray sky because of their extreme luminosity. About half of all observed x-ray sources belong to the AGN class.

For more information about these sources, their coordinates were compared with those of known extragalactic radio and infrared sources. This effort yielded almost 3000 AGNs that emitted radiation in the radio and x-ray ranges – about ten times more than had been known before ROSAT.

About 250 ROSAT sources were identified with infrared galaxies, of which more than half were known to be Seyfert galaxies. In addition, ROSAT found several infrared-intense spiral galaxies whose x-ray brightness by far surpasses the maximum recorded by the Einstein survey. The two brightest of these objects are classified as starburst galaxies, and there are indications that many others of this group show starburst activity.

X-ray astronomers received help in these identifications from their optical colleagues – active galaxies are easily recognized from their visible-light spectra. But since it would take too much time to obtain individual spectra of every galaxy in the sky in order to classify it, astronomers take images of an entire field of sky through a glass prism in front of the telescope. These objective prism plates can show only the brighter sources, but this selection effect is not too disturbing, since active galaxies are brighter than normal galaxies anyway.

The reliability of this method became apparent soon after the ROSAT survey was finished. A first comparison of positions of previously unidentified x-ray sources with possible AGNs on objective prism plates yielded 43 matches. Detailed individual spectra eliminated two of these objects as hot, bluish foreground stars. The remaining 41

Even in the far quasar 3C 273, at a distance of more than two billion light years, ROSAT observed x-rays from a jet of matter.

were predominantly classified as Seyfert galaxies or quasars, but a few turned out to be starburst galaxies or interacting pairs of galaxies.

Gigantic Energy Beacons

Radio and Seyfert galaxies had been identified as extragalactic sources from the beginning, but quasars originally looked to astronomers as if they were stars or at least star-like – the word *quasar* is short for "quasi-stellar radio source." The first quasars were discovered by radio astronomers in 1960, but without a very sharp view of the sky at the time, they could not provide exact positions. Optical astronomers had to search for possible counterparts, and when these were finally found, they looked unimpressive. But further research revealed peculiarities in the spectra of these candidates: The few lines visible in the spectra could not be attributed to any known element.

When the American astronomer Maarten Schmidt finally succeeded in providing an identification of the spectral lines in 1963, astronomers were faced with yet another problem: The spectra showed exceedingly high redshifts, which had to be interpreted in the framework of an expanding universe as correspondingly large distances. But that meant that quasars were more distant than all known galaxies of the time, and therefore they had to be much brighter as well to be visible from Earth. Some quasars shine brighter than a thousand galaxies. At the same time, intensity variations within a few months or even weeks made clear that the light-emitting region itself could not be very large – at most a few light weeks or months. How could these small objects produce so much radiation?

This question sparked a long-standing controversy that lingers even today. While the overwhelming majority of astronomers accept the cosmological interpretation of quasars as distant, and therefore extremely energetic, objects, a few cling to the interpretation of quasars being relatively close and thus "normal" with regard to their energy output. However, the proponents of this minority opinion have to invoke rather arcane scenarios to explain the observed redshift.

Today, the requirements for the central engine in a quasar do not seem that extreme. For 3C 273, one of the first quasars discovered, to sustain its luminosity of one hundred trillion stars (or about one thousand galaxies), a central black hole of one billion solar masses has to devour approximately a hundred stars of the size of our sun each year.

ROSAT can now be used to support the cosmological model for quasars. For many AGNs, ROSAT data clearly show an x-ray signature, which indicates a disk of matter in the immediate vicinity of the central energy source.

Clusters of Galaxies

Long before Edwin Hubble recognized the true size and distance of the Andromeda galaxy, observers had noticed unusual concentrations of nebulae, for instance in the direction of the constellation Virgo. The French comet hunter Charles Messier cataloged fourteen nebulous objects in this area within a diameter of only eight degrees.

The spectra of these galaxies show nearly

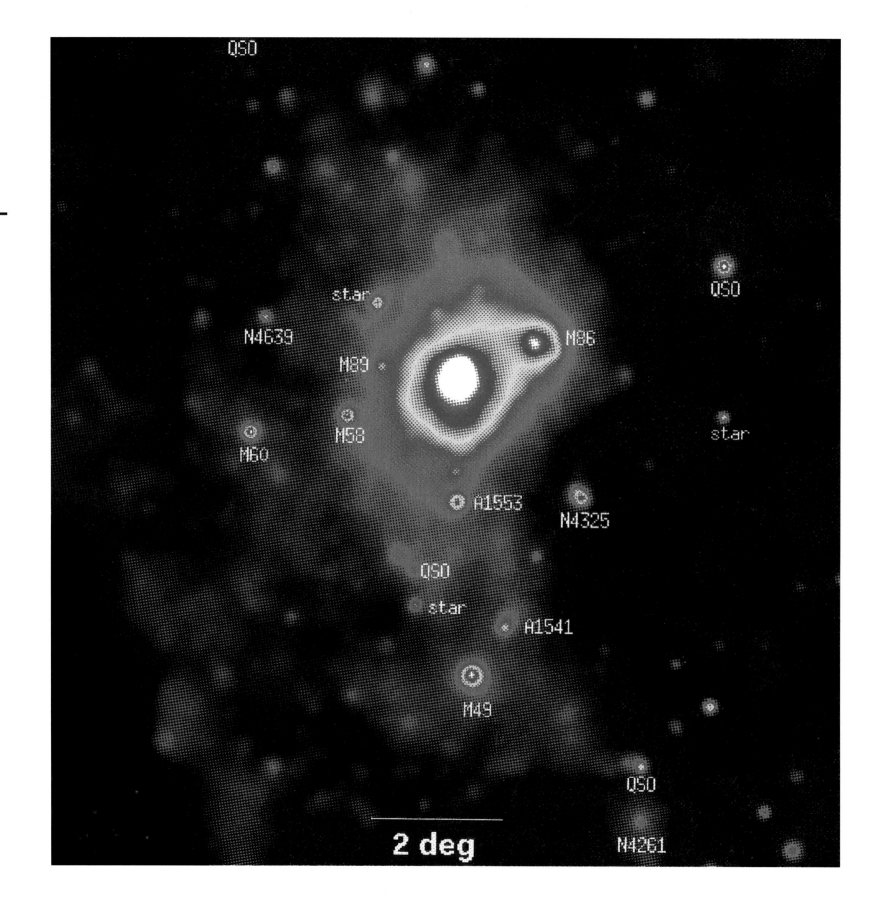

QSO

QSO

star

N4639

M86

M89

star

M60

M58

A1553

N4325

QSO

star

A1541

M49

QSO

2 deg

N4261

identical redshifts, meaning that together with hundreds of fainter objects in the same celestial region they had to be located at similar distances. They form a large accumulation of galaxies. This Virgo cluster is now known to be the closest cluster of galaxies; its center lies at a distance of about seventy million light years and coincides with the giant elliptical galaxy M 87. Today, astronomers know more than 6000 clusters of galaxies, and many of them feature an active galaxy in their center.

During a rocket experiment in March 1969, observations revealed x-ray emission from the direction of the Coma cluster, which is about 450 million light years away and does not have an active galaxy in its center. The radiation from this area was much more intense than expected from all members of the cluster together, so there had to be an additional source–perhaps thin, hot gas between the galaxies.

The existence of such intergalactic matter, occasionally called intracluster gas, had been suspected by radio astronomers. In many radio galaxies the extended radiating regions looked like trails of smoke from a chimney; this could only be explained by movement of the galaxy through surrounding matter. Apparently, galaxies in the crowded environment of a cluster rob each other of their extended halos, and their continuous movement within the cluster heats the intergalactic gas to x-ray-emitting temperatures.

During the 1930s, Fritz Zwicky established from spectra of the Coma galaxies that individual galaxies behaved like a swarm of insects, moving irregularly with respect to each other at speeds of several hundred to a thousand kilometers per second. Zwicky also noticed that the combined gravitational force of all Coma galaxies together would not be sufficient to keep the wildly moving galaxies together permanently. If the cluster was not just a transient phenomenon, it had to hold thirty times more matter than was contained in all the stars visible in these galaxies. This was the first, but then little noticed, indication of the existence of dark matter.

At the beginning of the 1970s, the Uhuru satellite confirmed the diffuse nature of the x-ray emission from the Coma cluster and identified other galaxy clusters as diffuse x-ray sources. From the energy distribution of this radiation, the temperature of the hot gas was calculated at between ten million and a hundred million degrees.

At the end of the seventies the Einstein satellite provided the first high-resolution x-ray views of galaxy clusters; astronomers were now able to study the distribution of the hot gas within the cluster. One of the results showed that galaxy clusters with a large dominant central galaxy also exhibit a strong concentration of the gas toward the center.

With its energy range, sensitivity, the resolution of its telescope, and its virtually unlimited field of view, ROSAT is particularly well suited to observe the diffuse x-ray emission from galaxy clusters. The distribution of the x-ray-bright gas within the Virgo cluster showed a surprisingly complex structure. The freely movable gas collects in areas of high gravitational potential (like rain in potholes of a road) and so provides insight into the inner

X-rays from ten-million-degree intergalactic gas reveal its complex structure in the Virgo cluster, showing at least one subcenter around the galaxies M 84 and M86 (right) in addition to the known center around the giant elliptical galaxy M 87.

structure and dynamic development of the cluster. The center (in the vicinity of M 87) reveals an almost spherical core region with continuously decreasing density toward the outside, indicating relatively old age.

Superimposed on this core region are several subcenters: in the east around the galaxy pair M 59 / M 60, in the west around the pair M 84 / M 86, and in the south around the galaxy M 49. Strong gravitation from the center of the cluster will make these subcenters merge with the core region in a few billion years, forming an old, "relaxed" cluster.

Similar substructures also have been observed in the Perseus cluster and in the Coma cluster. Both had been considered old, mature galaxy clusters, where an inner equilibrium had had time to develop. Their structured appearance in spite of their age can only mean that smaller groups or clusters of galaxies were added from the outside, increasing the cluster's mass over time.

From the temperature and distribution of the gas we can calculate not only the mass of the intergalactic matter within a given radius around the core but also the total mass of the cluster in this zone, as it is responsible for the observed distribution of the gas. This shows again that the hot intergalactic gas, detectable only in the x-ray range, contains four to five times more mass than the total of stars and interstellar gas and dust within the galaxies.

But even this mass is not sufficient to solve the dark matter problem initially posed by Zwicky. Even with this intergalactic gas we can account for only 20 to 25 percent of the matter making itself felt by gravitation.

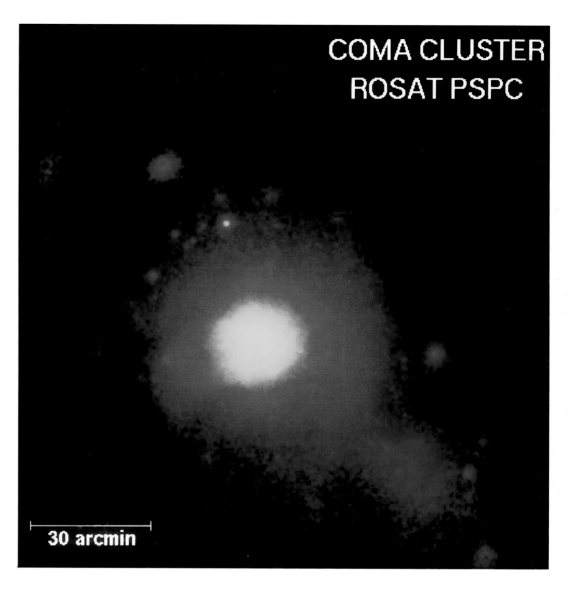

COMA CLUSTER
ROSAT PSPC

30 arcmin

A Deep View Through the Lockman Hole

The Italian scientist Galileo Galilei must have been quite surprised when he pointed his telescope at the Milky Way at the beginning of the seventeenth century. Instead of the diffuse, "milky" glow he saw many individual stars, too faint and too close to each other to be seen with the naked eye. History repeated itself – albeit with different protagonists – with ROSAT. Since the first successful observation of the x-ray sky in 1962, the discovery of the source

While the Coma cluster has long been considered an example of an old, quiescent cluster, the distribution of hot, x-ray-emitting intergalactic gas indicates the reality of a large cluster of galaxies merging with a smaller one.

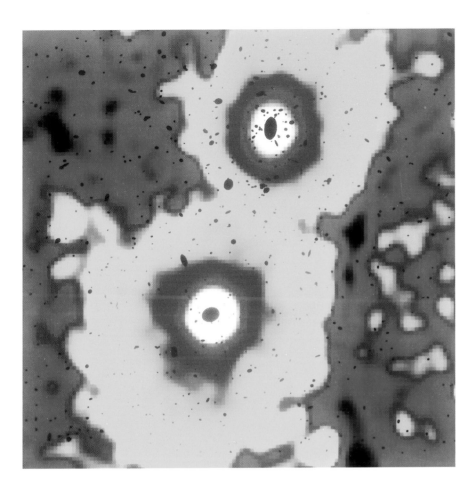

The galaxy cluster Abell 3528 shows a binary structure in the range of x-rays coming from the hot intergalactic gas; this is due to two subclusters that will have merged into one within a few hundred million years. The color-coded x-ray image has been overlaid with the positions and sizes of the optically visible galaxies.

The galaxy cluster Abell 754 is at an advanced state of merging of a small group of galaxies with the main cluster. The hot intergalactic gas reveals this process by the irregular distribution of x-ray intensities (indicated by color coding and contours) and by the corresponding temperature differences between forty and eighty million degrees.

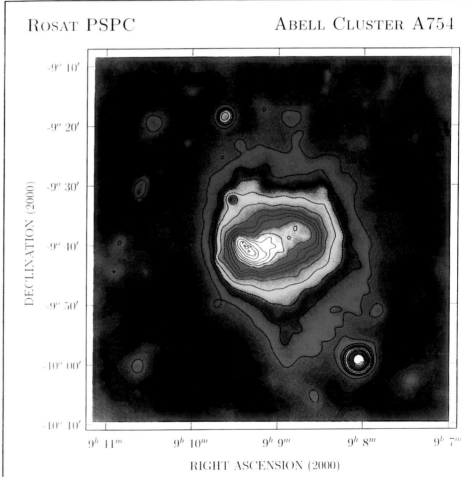

ROSAT PSPC   ABELL CLUSTER A754

DECLINATION (2000)

-9° 10'
-9° 20'
-9° 30'
-9° 40'
-9° 50'
-10° 00'
-10° 10'

$9^h 11^m$ $9^h 10^m$ $9^h 9^m$ $9^h 8^m$ $9^h 7^m$

RIGHT ASCENSION (2000)

Scorpius X-1, an apparently diffuse radiation component was known, which seemed to reach Earth from all directions more or less uniformly. The better resolution of its x-ray telescopes and the higher sensitivity of its detectors made it possible for the Einstein satellite to attribute about 15 percent of this emission to individual sources. ROSAT increased this number to more than 75 percent.

This became possible as part of an extremely long-term observation in a particular field of the sky called Lockman hole, where interstellar matter in our galaxy absorbs no x-rays. The long-term observation resulted from a superposition of numerous individual measurements, involving a total observing time of about eight days by May 1996. The inner part of the ROSAT field of view then contained about 250 discrete sources; this corresponds to about 720 sources per square degree, or almost 150 sources in an area the size of the full moon. In the meantime, the majority of sources were identified with optical and/or radio-astronomical objects. These observations indicate that the lion's share of this radiation emitted by discrete sources (about 80 to 85 percent) originates in active galactic nuclei, with the rest being contributed by galaxy clusters and single galaxies.

Scientists are measuring the redshift of these sources by optical means to obtain information about their distance. Then the individual sources can be arranged in space and – according to the motto "the farther away, the younger" – in time, so that the temporal development of the x-ray sources along the line of sight emerges. The average redshift of a sample of sources indicates that the light now reaching us left most of them when the universe was only one-fourth of its current age. The absolute age depends heavily on the value of the Hubble constant, which relates the measured redshift to the expansion of the universe. Since this value is still the subject of intense investigations and discussions, we will restrict ourselves to relative ages. In any case, all these sources are optically extremely faint objects, and for now their spectra can be obtained only with the Keck ten-meter telescope on Mauna Kea.

However, it is not easy to derive information on the temporal development of x-ray sources directly from the measured data, since they can be influenced by the curvature of space. The problem facing astronomers is like the task of determining the age of trees, given only the observation of an infinite row of trees and the knowledge that age decreases with distance until it somewhere reaches the point where the trees are just being planted.

As long as the growth of the trees can be disregarded – which is certainly true for the mature trees closest to us – the height of the trees decreases with distance in a measurable way, following the principle of "twice the distance, half the height." As we determine the number of trees as a function of their apparent size, we will find fewer large trees and increasingly more seemingly small ones. But if we reach the zone where the trees are still growing, the number of progressively smaller trees will increase much more slowly, simply because from this point onward no truly large trees exist.

The galaxy cluster Abell 2256 lies at a distance of about one billion light years and contains more than 500 galaxies. The x-ray signal comes from hot intergalactic gas with a temperature of approximately eighty million degrees. The binary structure in the distribution of the gas detected by ROSAT makes it clear that two galaxy clusters of different sizes are merging.

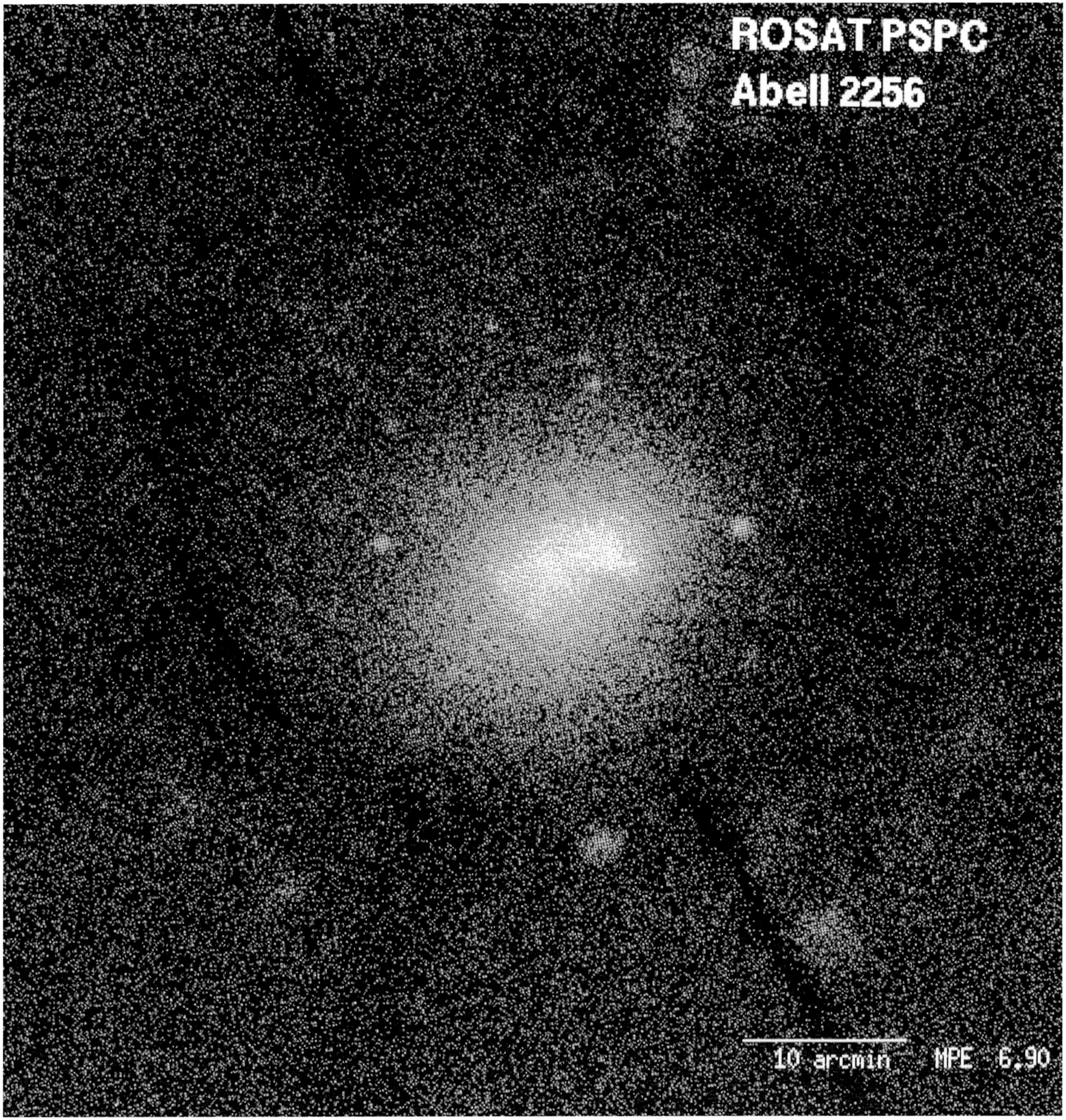
ROSAT PSPC
Abell 2256
10 arcmin
MPE 6.90

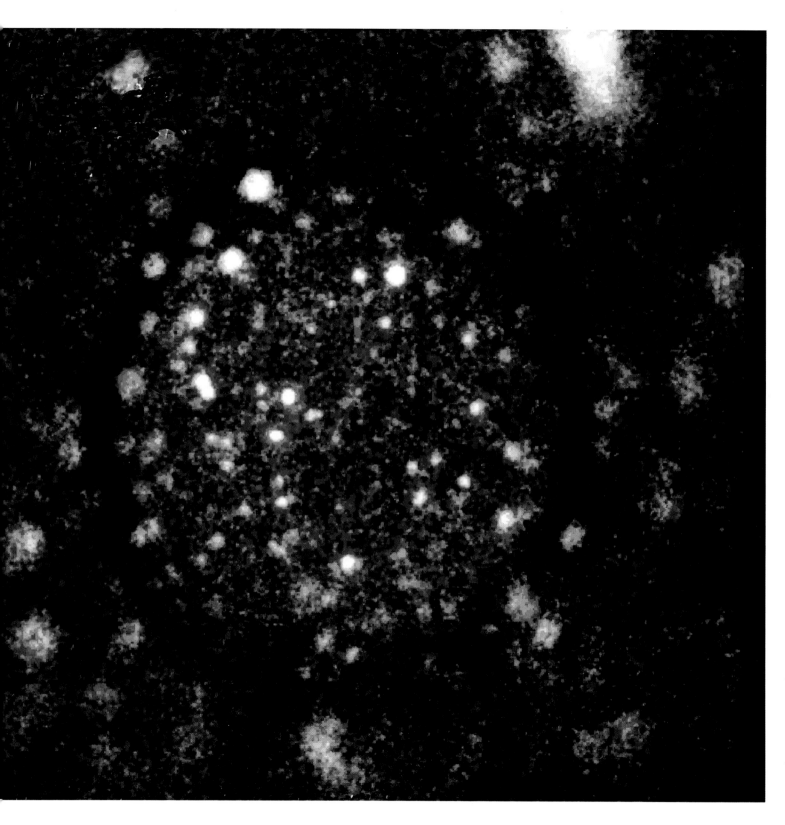

This ROSAT image, exposed for more than 53 hours, shows an area of the sky in the Big Dipper, known as the Lockman hole. It is the deepest x-ray image of the sky up to now. It shows more than one hundred sources per square degree, most of them extragalactic. There are soft (red), medium-hard (green), and hard (blue) x-ray sources. The objects with a hard x-ray spectrum are probably active galaxies. This image proves that at least 75 percent of the seemingly diffuse x-ray background radiation comes from overlapping extragalactic point sources, which could not be resolved in the pre-ROSAT era.

This consideration, however, is only applicable to infinite, flat surfaces. If we take the curvature of Earth into account, the number of apparently smaller trees close to the horizon decreases disproportionately quickly, since even grown distant trees extend only partway above the horizon. Farther away they become fully invisible.

The trees of x-ray astronomers are x-ray-active galactic nuclei and quasars; their distance can be determined spectroscopically up to a limit determined by the sensitivity of the measuring devices. Today's investigations indicate a more or less uniform increase in the total number of sources up to a certain brightness limit – within this limit, however, a potential curvature of space is difficult to detect. Beyond this limit the increase slows down; this could be an effect of space curvature, or it could be related to the development of the sources (the growth of the trees), or both. Only when we succeed in obtaining even more far-reaching observations and in understanding the temporal development of quasars will we be able to account for the developmental effects in the data so that only the influence of space curvature remains.

Epilogue

Since its launch in June 1990, ROSAT has made history. Within the first half year it increased the number of known x-ray sources from 5000 to almost 70,000, and since then it has added about as many more. Demand for observing time remains undiminished; there are four times more proposals for observing time than can be accommodated. But even the most successful satellite will ultimately only be history when vital parts of the satellite systems or the scientific payload cease to work.

National and international successor projects are underway to allow the continuation of x-ray astronomy after ROSAT. Some of them are already operational; others will become reality within a decade.

For some time, several x-ray satellites have been working in parallel to ROSAT, for instance the Russian Granat or the Japanese ASCA, built with substantial American participation. While Granat has observed hard x-ray emission (between 35 and 250 keV) since the beginning of 1990, ASCA – launched in 1993 – features x-ray CCD cameras with high energy resolution between 0.5 and 10 keV, making it possible to analyze discrete x-ray emission lines. ASCA data, for instance, have contributed significantly to the identification of conspicuous areas of radiation just outside the Vela SNR as debris from the original star, based on their chemical composition. Both satellites, however, have very limited spatial resolution in comparison to ROSAT.

The next two large x-ray projects are American and European initiatives. The American one, the Advanced X-ray Astrophysics Facility (AXAF), is expected to be launched by the end of 1998.

AXAF features a large x-ray telescope with focal length of 10 meters (33 feet) and an aperture of 120 centimeters (4 feet), consisting of four concentric Wolter mirror shells; with its microchannel detector, similar to the HRI aboard ROSAT, it should reach a resolution of 0.5 arc seconds, ten times better than ROSAT. An x-ray CCD is planned as a second type of detector. High-resolution spectra will be obtained with the help of special transmission gratings, developed in cooperation with the Max Planck Institute for Extraterrestrial Physics and the Laboratory for Space Research in Utrecht, the Netherlands.

AXAF's counterpart is the European Space Agency's X-ray Multi-Mirror satellite (XMM). It should be launched in 1999, equipped with three large Wolter mirror systems, each with a focal length of 7.5 meters (25 feet) and consisting of 58 concentric mirror shells. Two of the telescopes will be equipped with CCDs from the University of Leicester, England, and with a reflecting spectrometer from Utrecht; the third one will focus onto a new type of x-ray CCD detector with an effective area of six by six centimeters (2.4 by 2.4 inches). It differs from other CCDs in having a large efficiency at high energies (up to 15 keV) and very good time resolution, so that it is particularly suited to observing short-term intensity variations. This detector has been developed by the MPE and has been manufactured in a laboratory built for this project in cooperation with the Max Planck Institute for Physics in Munich. The scientific guidance for development and production of the x-ray mirror systems rests with the MPE.

As part of the national program, a second German x-ray satellite, called ABRIXAS (A BRoad-

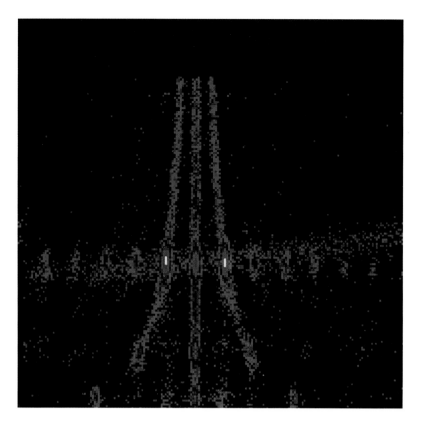

band Imaging X-ray All-sky Survey), will be launched in 1999 as well. It will survey the sky with an imaging x-ray telescope in the range between 0.5 and 15 keV for the first time, extending the ROSAT survey into shorter wavelengths. This will bring new insights in areas where ROSAT's x-ray range was limited by interstellar absorption, for instance in observations of active galactic nuclei hidden within dense accretion disks.

ABRIXAS will carry a six by six centimeter CCD detector identical to that used on XMM. Seven Wolter telescopes containing 27 concentric mirror shells each will scan the entire sky and focus x-rays onto this detector. ABRIXAS will thus provide an all-sky survey for XMM and AXAF, as ROSAT did for its own mission during its first six months.

XMM will provide mainly detailed and highly accurate x-ray spectra, thanks to its large collecting area. This will aid the investigation of galaxy clusters by helping determine the spatial distribution of temperature and element abundances in the intergalactic gas. The comparison of gal-axy clusters at different distances – and therefore of different ages – should provide clues about the chemical evolution of the universe.

XMM and AXAF will make it possible to study neutron stars in detail, for instance by studying the composition of their surfaces. This will finally provide information about this third category of stars – after "normal" stars, where the ordinary gas and radiation pressure counter the inwardly directed gravitational force; and white dwarfs, which are stabilized by the pressure of the degenerate electron gas. The pressure of neutron gas is what stabilizes neutron stars.

When theoreticians indicated for the first time the possible existence of neutron stars, the Russian physicist Lev Landau called them "ghost stars," as he thought they would be too small and too faint to be discovered or even investigated over cosmic distances. But ROSAT has observed about two dozen neutron stars in the x-ray range, one of them in the constellation Sagittarius, which possibly wanders our galaxy as an old, solitary object: This source is at least 7000 times brighter in the x-ray range than in the optical part of the spectrum, and indeed an optical counterpart has not yet been identified. Such a high ratio of x-ray to optical luminosity is predicted by theories of the radiative behavior of old, solitary neutron stars.

Finally, XMM and AXAF will try to obtain more information about dark matter. It must exist in galaxies and galaxy clusters, since its additional gravitational force is required to explain the observed rotation curves of galaxies and the stability of galaxy clusters – the gravitation of the visible matter alone is insufficient. From careful observations of the hot gas in galaxies and galaxy

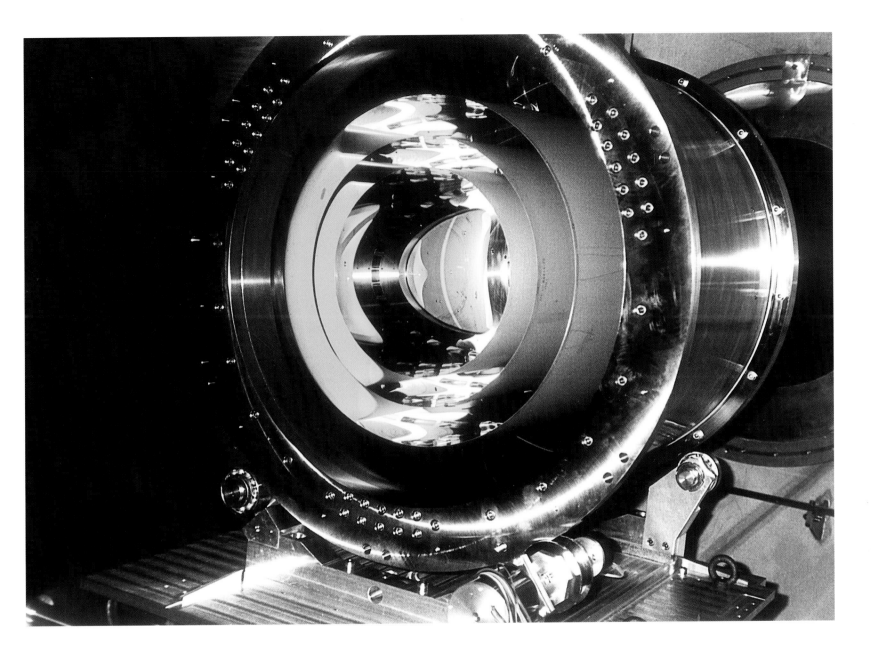

clusters the structure of the effective gravitational fields will be determined, leading in turn to the derivation of mass and distribution of dark matter.

Although dark matter constitutes most of the total mass of the universe, for now it evades detection. Its exploration could open the door to yet another invisible sky – a sky that has remained invisible even to ROSAT.

List of Acronyms

AGN	Active Galactic Nucleus
AS&E	American Science and Engineering Company
ASCA	Advanced Satellite for Cosmology and Astrophysics
AXAF	Advanced X-ray Astronomy Facility
CCD	Charge Coupled Device
DARA	German Agency for Space Affairs (Deutsche Agentur für Raumfahrtangelegenheiten)
DFG	German Science Foundation (Deutsche Forschungsgemeinschaft)
DLR	German Research Institute for Aeronautics and Space Flight (Deutsche Forschungsanstalt für Luft- und Raumfahrt)
ESA	European Space Agency
eV	Electron Volt
EXOSAT	European X-Ray Observatory Satellite
GeV	Giga-Electron Volt
GSOC	German Space Operations Center
HD	Henry Draper Catalogue
HEAO	High-Energy Astronomical Observatory
HEXE	High-Energy X-Ray Experiment
HRD	Hertzsprung–Russell Diagram
HRI	High-Resolution Imager
IC	Index Catalogue
KASCADE	Karlsruhe Air Shower Core and Array Detector
keV	Kilo-Electron Volt
LMC	Large Magellanic Cloud
M	Messier Catalogue
MBB	Messerschmidt–Bölkov–Blohm
MeV	Mega-Electron Volt
MIT	Massachusetts Institute of Technology
MPE	Max Planck Institute for Extraterrestrial Physics (Max-Planck-Institut für Extraterrestrische Physik)
NASA	National Aeronautics and Space Administration
NGC	New General Catalogue
PSPC	Position-Sensitive Proportional Counter
PSR	Pulsar
SAS	Small Astronomy Satellite
SMC	Small Magellanic Cloud
SS	Stephenson–Sanduleak (catalog)
XMM	X-ray Multi-Mirror satellite

Index